Systems Biology

Edited by Dimitrios Vlachakis

Published in London, United Kingdom

IntechOpen

Supporting open minds since 2005

Systems Biology
http://dx.doi.org/10.5772/intechopen.79601
Edited by Dimitrios Vlachakis

Contributors
Juan C Ramirez, Jana Žiarovská, Lucia Zeleňáková, Shuhong Luo, Ruixue Wang, Pilar Garcia-Jimenez, Rafael Robaina, Holger Husi, Marco Fernandes, Dimitrios Vlachakis, Eleni Papakonstantinou, Katerina Pierouli, Elias Eliopoulos

First published in London, United Kingdom, 2019 by IntechOpen
IntechOpen is the global imprint of INTECHOPEN LIMITED, registered in England and Wales, registration number: 11086078, The Shard, 25th floor, 32 London Bridge Street
London, SE19SG - United Kingdom
Printed in Croatia

British Library Cataloguing-in-Publication Data
A catalogue record for this book is available from the British Library

Additional hard and PDF copies can be obtained from orders@intechopen.com

Systems Biology
Edited by Dimitrios Vlachakis
p. cm.
Print ISBN 978-1-83880-803-7
Online ISBN 978-1-83880-804-4
eBook (PDF) ISBN 978-1-83880-805-1

For EU product safety concerns:
IN TECH d.o.o., Prolaz Marije Krucifikse Kozulić 3, 51000 Rijeka, Croatia, info@intechopen.com or visit our website at intechopen.com.

Meet the editor

Handling, managing and analyzing big biomedical data from a repertoire of different biological and physical disciplines are huge challenges in modern science. A few years ago, techniques were being developed in an effort to generate high-quality artificial datasets because information was scarce. Statistical viability was almost impossible with only a dozen genes and a few proteins available to analyze. However, after the human genome project was completed, data accumulation sky-rocketed. A couple of decades later, data storage became more expensive than data analysis. Systems biology is a modern field of science, attempting the impossible: modeling complex biological systems in a holistic way, by using only the important bits of information and removing the noisy parts. The chapters of this book provide insights into a range of modern applications and breakthroughs in the field of systems biology. All chapters have been carefully selected, adjusted, and fine-tuned in a seamless way that helps them achieve synergy and makes it easier for both the novice and the expert reader to follow. A lot of effort has gone into providing the scientific basis of complex biological systems, since combining and fusing information from different -omics domains are not trivial. Different experiments under different conditions cannot be easily fused without the risk of increasing the noise in the fused dataset. In this book, a set of quite diverse chapters has been carefully put together to give the reader a broad overview of today's systems biology domain.

I would like to close this preface with the words of J.R.R. Tolkien, who stated: "All that is gold does not glitter, Not all those who wander are lost." More than 98% of our genome is uncharted and things are even worse in other -omics domains. The era of the dark genome is upon us. Systems biology is attempting to fuse huge datasets from only a tiny portion of the full complex biomedical spectrum in a meaningful and reproducible way, so that science can evolve, and new and validated knowledge can be attained.

Contents

Preface

Handling, managing and analyzing big biomedical data from a repertoire of different biological and physical disciplines are huge challenges in modern science. A few years ago, techniques were being developed in an effort to generate high-quality artificial datasets because information was scarce. Statistical viability was almost impossible with only a dozen genes and a few proteins available to analyze. However, after the human genome project was completed, data accumulation sky-rocketed. A couple of decades later, data storage became more expensive than data analysis. Systems biology is a modern field of science, attempting the impossible: modeling complex biological systems in a holistic way, by using only the important bits of information and removing the noisy parts.

The chapters of this book provide insights into a range of modern applications and breakthroughs in the field of systems biology. All chapters have been carefully selected, adjusted, and fine-tuned in a seamless way that helps them achieve synergy and makes it easier for both the novice and the expert reader to follow. A lot of effort has gone into providing the scientific basis of complex biological systems, since combining and fusing information from different -omics domains are not trivial. Different experiments under different conditions cannot be easily fused without the risk of increasing the noise in the fused dataset. In this book, a set of quite diverse chapters has been carefully put together to give the reader a broad overview of today's systems biology domain.

I would like to close this preface with the words of J.R.R. Tolkien, who stated: "All that is gold does not glitter, Not all those who wander are lost." More than 98% of our genome is uncharted and things are even worse in other -omics domains. The era of the dark genome is upon us. Systems biology is attempting to fuse huge datasets from only a tiny portion of the full complex biomedical spectrum in a meaningful and reproducible way, so that science can evolve, and new and validated knowledge can be attained.

Dimitrios P. Vlachakis
Professor,
Agricultural University of Athens,
Athens, Greece

Section 1

Introduction

Chapter 1

Introductory Chapter: Systems Biology Consolidating State of the Art Genetics and Bioinformatics

Eleni Papakonstantinou, Katerina Pierouli, Elias Eliopoulos and Dimitrios Vlachakis

1. Biological systems and computational tools in systems biology

A biological system is defined as the network of biological entities with a specific function. Although the study of the genome and its expression process remains important, in recent years, researchers have focused on studying the structure and dynamics of an entire biological system. The operation of a system determines the interactions of its individual elements. Understanding the function and interactions of a system is accomplished by comprehension of four basic attributes: (1) structures of the system, including the gene interactions and biochemical pathways, as well as the control mechanisms of the above procedures, (2) system dynamics, containing patterns of the system behavior over time, and the identification of the main mechanisms that control specific behaviors under various conditions, (3) the control method, including configuration of cellular mechanisms in order to reduce dysfunctions, (4) the design method, including techniques for designing and modifying biological systems to reduce errors. The definition of systems biology has not yet been clarified; many different definitions are depending on the researcher. In general, systems biology is an interdisciplinary field of biology that involves the computational and mathematical modeling of complex biological systems. The purpose of this field is to understand the complex interactions and functions at the organism, tissue or cell level with direct application in biomedical research. Therefore, systems biology comprises an approach that diverges from heretofore reductive biological research.

Systems biology examines the interactions between several components rather than the individual features of the molecules, in order to understand the phenotype resulting from the components of the system. To this end, computational approaches are employed in systems biology to create possible *in silico* models that can also be verified experimentally *in vivo* or *in vitro*, thus allowing the analysis of a large number of data [1–5]. In the study of biological systems, various computational tools are used including techniques for sequence alignment and for recording molecular dynamics, molecular interactions and discovering or predicting the molecular structure [6].

The computational techniques are divided into two categories; those guided by data and those guided by hypothesis. Network modeling is based on data to provide information on the interactions of numerous molecular components. On the other hand, dynamic modeling is based on a hypothesis in order to characterize the quantitative relations and interactions between the molecular components (**Figure 1**).

This category of computational techniques includes strategies that involve smaller biological systems with fewer molecular components [7].

In the context of the computational models' employment in systems biology, there is a correlation between the design of a biological model and the development of a computer program. Thus, the researcher can visualize the internal processes and flows of system reactions, as well as to study the changes that occur depending on the conditions of the system [8].

In a biological network in systems biology, the molecular components are considered as "nodes", while their interactions as "edges". Molecular components include genes, proteins, metabolites, drugs, and can also be diseases and phenotypes [9]. On the other hand, the dynamical modeling of a biological system involves the representation of known molecular paths in a mathematical form. Mathematical equations are intended to describe biological processes based on the physicochemical properties that determine the rate of reaction and the affinity of the interaction. In order to achieve the most efficient performance of a dynamic model, the initial concentrations of cellular components must be given, and their connectivity is then established. In this way, through dynamic modeling, the concentrations of the components can be calculated at subsequent times and can be compared with those that occur in experimental time [5]. A dynamic model can be definitive when the initial data and parameters are determined and lead to a particular path or can be stochastic when it is able to proceed to different conditions with different probabilities according to its primary data [10].

Various tools have been developed for analyzing and modeling biological systems. An indicative example is the biological tool COPASI, which is suitable for mathematical simulation and analysis of the dynamics of biochemical processes [11].

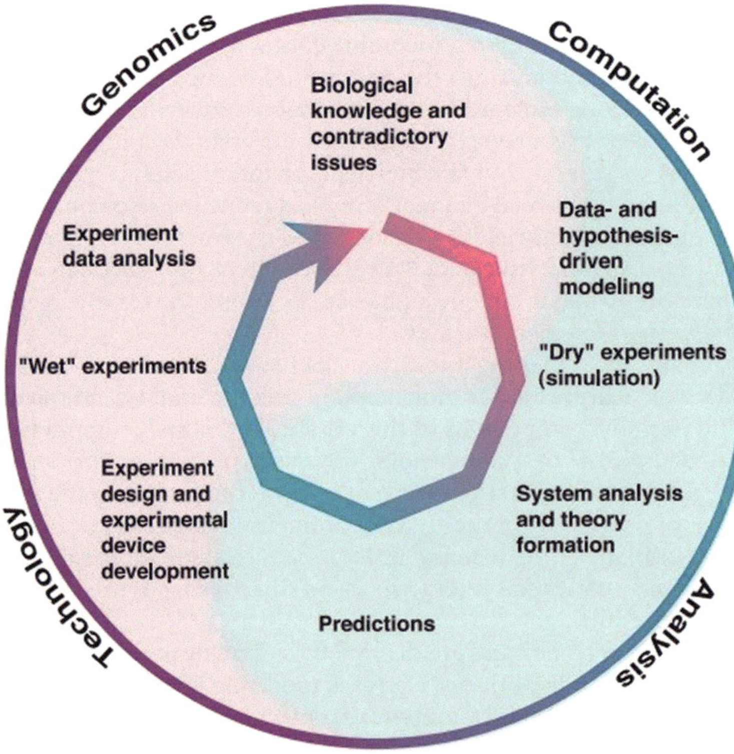

Figure 1.
Hypothesis-driven research in systems biology [2].

BioModel Analyzer (BMA) is a graphical tool for the structure and analysis of biological systems. It aims to combine the simple analysis offered by formal-verification tools to the biologists' expectations of examining systems spatially and temporally and studying the modifications of system interactions according to several conditions [12].

BioDiVinE is a tool for parallel study and analysis of biological systems. The tool analyzes the model based on its chemical reactions, through the following main features: (a) the specificity of models according to their chemical equations through the user interface, (b) the representation of models based on the various ordinary differential equations (ODEs), (c) adjustment of the initial kinetic parameters and conditions, (d) regulation of the discrete abstraction parameters, (e) graphic simulation of model's discrete stages and (f) model controlling [13].

Another example is BoolNet, a computational tool which studies the biological Boolean networks (BNs) that describe the gene regulation where the genes exhibit dual "on/off" behavior, i.e., whether they are expressed or not. The tool can analyze new functions for attractor search, robustness and binarization for three categories of BNs: (i) synchronous BNs, consisting of a set of genes (variables) and a set of transition functions, (ii) asynchronous BNs, where a single transition function is selected, and the corresponding variable is updated, and (iii) probabilistic BNs, that allow the specification of various transition functions per variable [14].

2. Applications of systems biology

While traditional reductionist methods provide an invaluable insight into the molecular mechanisms of a disease, a systems-level approach has been widely used in recent years based on the integration of large-scale data emerging from omics technologies. The acquired high-dimensional data produced by genomic, transcriptomic, proteomic, lipidomic and metabolomic technologies can be processed and analyzed by employing novel techniques in data analysis and data mining, bioinformatics and machine learning approaches. The ultimate goal is the design of a refined computational model that can reflect on the dynamic perturbations of the system in a predictive mode.

Systems biology approaches have been implemented in the study of many diseases, such as cancer and neurodegenerative diseases. Carcinogenesis, tumor progression, and metastasis have long been a challenging field of research, since high complexity interaction networks govern them at a genetic, epigenetic, cellular, tissue and environmental level [15]. The wide availability of high-throughput genomic, epigenomic and transcriptomic data on different types of cancer in large repositories, such as "The Cancer Genome Atlas (TCGA)," has empowered research in the context of cancer systems biology. The use of such data in parallel with computational models has been employed in a number of successful stories, identifying new key regulators, establishing predictive biomarkers and designing optimized or novel therapeutic strategies against cancer [16–19].

Another application of systems biology includes the computational analysis on extensive experimental data in the field of pharmacology, namely systems pharmacology. Systems pharmacology is focused on the study of drugs, identifying new drug targets, repurposing of existing drugs and analyzing the properties and effects of known drugs in a systems-level. Addressing the complexity of the cellular networks and the mode of action of a drug can lead to a better understanding of side effects and adverse events of a drug and the identification of off-targets, improving the safety and effectiveness of disease treatment [20]. In the past decade, systems-based applications have proved to gain better insights into drug-drug interactions, drug-target networks, drug-target interactions, and drug side-effects, leading to novel drug discovery [21–23].

Finally, in the field of virology, a systems-level approach is required to address the extremely complex viral and host processes that follow upon viral infection. Systems virology encompasses the study of virus-host interaction networks through the process and analysis of high-throughput data, such as next-generation sequencing, mass spectrometry, microarray technologies, and protein-protein interactions, aiming towards the generation of a representative dynamic model of virus-host interactions [24]. Such methods may reveal key network components that can be used as potential targets during viral infection and replication and lead to novel preventive or therapeutic antiviral strategies [25].

Summarizing, systems biology is a field that aims to analyze and model the molecular pathways and interactions of a broader set of several molecular components. The study of a system can be achieved by computational tools that enable scientists to visualize and record the progress of functions in a system over time. The ultimate ambition is the prediction of potential changes and interactions according to the dynamic conditions, making it possible to lead to the development of more effective therapies and handling several diseases.

Author details

Eleni Papakonstantinou, Katerina Pierouli, Elias Eliopoulos and Dimitrios Vlachakis*
Department of Biotechnology, Genetics and Computational Biology Group, Laboratory of Genetics, Agricultural University of Athens, Athens, Greece

*Address all correspondence to: dimvl@aua.gr

References

[1] Hillmer RA. Systems biology for biologists. PLoS Pathogens. 2015;**11**(5):e1004786

[2] Kitano H. Systems biology: A brief overview. Science. 2002;**295**(5560):1662-1664

[3] Wanjek C. Systems biology as defined by NIH. National Institutes of Health. 2011;**19**(6):1-10

[4] Bartocci E, Lio P. Computational modeling, formal analysis, and tools for systems biology. PLoS Computational Biology. 2016;**12**(1):e1004591

[5] Zou J et al. Advanced systems biology methods in drug discovery and translational biomedicine. BioMed Research International. 2013;**2013**:742835

[6] Nobile MS et al. Graphics processing units in bioinformatics, computational biology and systems biology. Briefings in Bioinformatics. 2017;**18**(5):870-885

[7] Faratian D et al. Systems pathology—Taking molecular pathology into a new dimension. Nature Reviews. Clinical Oncology. 2009;**6**(8):455-464

[8] Gay S, Soliman S, Fages F. A graphical method for reducing and relating models in systems biology. Bioinformatics. 2010;**26**(18):i575-i581

[9] Maayan A. Introduction to network analysis in systems biology. Science Signaling. 2011;**4**(190):tr5

[10] Sobie EA et al. Systems biology—Biomedical modeling. Science Signaling. 2011;**4**(190):tr2

[11] Hoops S et al. Copasi—A complex pathway simulator. Bioinformatics. 2006;**22**(24):3067-3074

[12] Benque D, Bourton S, Cockerton C, Cook B, Fisher J, Ishtiaq S, et al. Bma: Visual Tool for Modeling and Analyzing Biological Networks. Heidelberg: Springer-Verlag; 2012. pp. 686-692. 7358 of LNCS. Berlin

[13] Barnat J, Brim L, Cern I, Drazan S, et al. BioDiVinE: A framework for parallel analysis of biological models. Electronic Proceedings in Theoretical Computer Science. 2009;**9**:31-45

[14] Mussel C, Hopfensitz M, Kestler HA. BoolNet—An R package for generation, reconstruction and analysis of Boolean networks. Bioinformatics. 2010;**26**(10):1378-1380

[15] Du W, Elemento O. Cancer systems biology: Embracing complexity to develop better anticancer therapeutic strategies. Oncogene. 2015;**34**(25):3215-3225

[16] Agrawal P et al. A systems biology approach identifies FUT8 as a driver of melanoma metastasis. Cancer Cell. 2017;**31**(6):804-819 e7

[17] Chakraborty S et al. Onco-multi-OMICS approach: A new frontier in cancer research. BioMed Research International. 2018;**2018**:9836256

[18] Selvarajoo K. A systems biology approach to overcome TRAIL resistance in cancer treatment. Progress in Biophysics and Molecular Biology. 2017;**128**:142-154

[19] Walsh LA et al. An integrated systems biology approach identifies TRIM25 as a key determinant of breast cancer metastasis. Cell Reports. 2017;**20**(7):1623-1640

[20] Berger SI, Iyengar R. Role of systems pharmacology in understanding drug adverse events. Wiley Interdisciplinary Reviews. Systems Biology and Medicine. 2011;**3**(2):129-135

[21] Li P et al. Transcriptome inference and systems approaches to polypharmacology and drug discovery in herbal medicine. Journal of Ethnopharmacology. 2017;**195**:127-136

[22] Perez-Nueno VI. Using quantitative systems pharmacology for novel drug discovery. Expert Opinion on Drug Discovery. 2015;**10**(12):1315-1331

[23] Wu Z et al. Network-based methods for prediction of drug-target interactions. Frontiers in Pharmacology. 2018;**9**:1134

[24] Law GL et al. Systems virology: Host-directed approaches to viral pathogenesis and drug targeting. Nature Reviews. Microbiology. 2013;**11**(7):455-466

[25] Petit MJ, Shah PS. Mapping Arbovirus-Vector interactions using systems biology techniques. Frontiers in Cellular and Infection Microbiology. 2018;**8**:440

Section 2

Integrative Systems Biology

Chapter 2

Integrative Systems Biology Resources and Approaches in Disease Analytics

Marco Fernandes and Holger Husi

Abstract

Currently, our analytical competences are struggling to keep-up the pace of in-deep analysis of all generated large-scale data resultant of high-throughput omics platforms. While, a substantial effort was spent on methods enhancement regarding technical aspects across many detection omics platforms, the development of integrative downstream approaches is still challenging. Systems biology has an immense applicability in the biomedical and pharmacological areas since the main goal of those focuses in the translation of measured outputs into potential markers of a Human ailment and/or to provide new compound leads for drug discovery. This approach would become more straightforward and realistic to use in standard analysis workflows if the collation of all available information of every component of a biological system was ensured into a single database framework, instead of search and fetch a single component at time across a scatter of databases resources. Here, we will describe several database resources, standalone and web-based tools applied in disease analytics workflows based in data-driven integration of outputs of multi-omic detection platforms.

Keywords: systems medicine, bioinformatics, omics, data integration, pathway analysis

1. Introduction

Over the last decade the emergence of high-throughput screening platforms and the increase in availability of large-scale-omics data, as well as clinical data from electronic health records comprising phenotypic, therapeutic and environmental factors information opened the possibility to mechanistically understand diseases and diseases stages at the molecular level. Thereby, a great number of wealth data in many kidney and cardiovascular conditions was generated, however these findings were neither translated nor reached the clinical setting and are still enclosed in peer-reviewed literature and across general scope expression profiling databases. Simultaneously it has become apparent that the existing systems to integrate and correlate this data are either inadequate or non-existent. Due to the multi-factorial molecular phenotype of disease, it is evident that development of novel therapeutic and disease detection approaches should be based upon the study of the entire "System" simultaneously. **Figure 1** gives a general overview in the fundamental difference between conventional and systems approaches, whereby in the context of conventional approaches a hypothesis is put forward that is assumed to be of importance in the disease or biological condition. This hypothesis is then tested

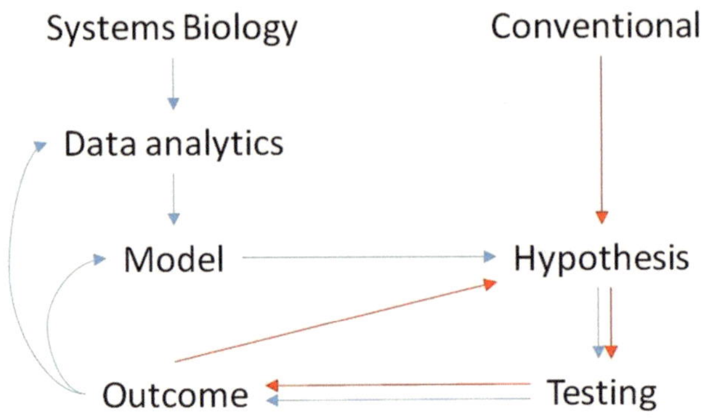

Figure 1.
Overview of general differences between conventional and Systems Biology approaches in biological and disease analysis research. Red arrows show the path of the conventional hypothesis-driven methodology including testing of a hypothesis, usually employing lab-based investigations, and re-adjusting the hypothesis dependent on outcomes. Blue arrows denote a systems approach, where data are integrated and analysed, producing a model system and a hypothesis that can be verified using conventional methods. Outputs of such an approach are usually fed back into the model or the data analysis stream to refine models, adjust hypothesis or confirm the established model.

and either validated or refuted based on the outcome of this hypothesis-driven methodology. Yet, it is obvious that it is easy to investigate any hypothesis and then choose the one that appears most correct, in the real world constraints such as time and financial resources do not allow for such an approach, and hypotheses are usually generated on a best-guess basis which can lead to a substantial amount of bias, resulting in skewed or partial insights and can often be misleading. In order to avoid such scenarios, research driven by the data itself rather than a hypothesis has been proposed a long time ago, but could not be properly implemented due to the lack of unbiased large-scale data or the ability to integrate disparate data in the first place. Additionally, a successful systems approach requires underlying prior knowledge, such as physicochemical parameters in how molecules interact with each other, what reactions they are involved in and other unconnected information. This knowledge has only slowly been accumulated through conventional research and has only over the last 10–15 years been available to such an extent where a systems approach became feasible. Data-driven systems biology-based diagnostic and prognostic models consisting of relevant panels of molecules—key branches of the cellular network, appear to more accurately reflect pathophysiology than traditional hypothesis-driven approaches, consequently, may have a much higher chance of success and implementation in the clinical setting. Of the most pronounced effects is the crossing between research borders and the urge for multidisciplinary integration of biology, chemistry, computing sciences, mathematics, and medicine to tackle the complexity of such system. To get a holistic view of a system's biology, multiple and different types of observations must be combined, such as clinical which includes pathological, demographical, epidemiological, and as well as molecular, which includes large-scale genotyping, gene expression, proteomics, metabolomics, and lipidomics data. The downside of such an approach in disease analytics or data integration is the rise in complexity both in output as well as in methods needed to generate those, and the skills required to interpret and contextualise outcome parameters. However, biological and disease models generated this way allow for a higher confidence in generating testable hypotheses, disease classifications on a molecular level and identification of overlapping and divergent pathways of malignant conditions. Ultimately, the removal of bias and integration of all available data, both clinical and biological, leads to a far better understanding of disease and enables the identification of intervention points with higher confidence and accuracy.

2. Disease classification boundaries

The standard resource for disease taxonomy relies primarily on the International Classification of Diseases (ICD) which displays information on diseases and health conditions, and a continuous monitoring of the associated epidemiological statistical trends World Health Organisation [1]. The foundations of the ICD disease classification relies mainly in a type of evidence-based medicine with distinction of clinical features, including patient symptoms, histological assessment, and evaluation of risk factors [2]. While widely used in the clinical setting, in the era of "big-data" and precision medicine, its rigid hierarchical structure lacks the flexibility needed to accommodate the fast and expanding molecular-insights of disease-phenotypes captured across many -omics platforms [3]. Moreover, to support this notion of undefined disease boundaries across current disease classification, we can observe the existence of co-occurring conditions that if seen as a unified biological network, could provide information about common multi-functional genes, cellular pathways, as well the impact of lifestyle [4]. Additionally, analysis of disease progression with the presence of overlapping conditions through evaluation of temporal correlation and disease progression patterns condensed from a population can become useful in the prediction and prevention at the patient's individual level in future disease-associate events [5].

Further disease taxonomy refinement can be achieved by applying network analysis [3] of combined disease phenotypes sourced from ICD-9 with protein-protein interactions (PPI's) data from STRING [6] and additional curation efforts of gene-disease associations (GDA) from several data sources. The network analysis allowed for reclassified of pancreatic cancer into 11 subclasses, which is consistent with the number of molecular subtypes observed in the Bailey et al. [7] study. They also proposed the use of such approach in drug repurposing, for instance therapy with metformin, a well-known agent used to treat type 2 diabetes mellitus (T2DM), that could regulate the imbalanced status of the microbiota community in the gut mucosae, a known cause of pathological chronic bowel inflammation as occurs in Crohn's disease and ulcerative colitis [8], and also act as preventable agent to reduce the risk of colorectal cancer. Moreover, molecular profiling associated with histologic assessment seems to yield enhanced probabilistic scores in graft survival predictions. For instance, joint integration of multi-center histology features in renal biopsies and gene-array data yielded a new molecular score system able to predict renal graft survival [9] and improving the diagnosis of antibody-mediated rejection of transplanted in hearts [10]. Such approaches can also be implemented to assess disease trajectory, treatment selection and monitoring in many neoplasms, and could be specially tailored for cases where the tumour primary site is of unknown origin [11].

3. Systems biology towards systems medicine

Over the last 15 years, the rise of systems biology as a research field has changed how we look at human normal physiological function and has helped to uncover disease complexity. Now scientists use systems biology approaches to understand the big picture of how all the pieces interact in an organism. The inference of genotype-phenotype relationships boosted by the assembly of a high-quality human genome opened the avenue for the development of reference maps of interactome networks, [12] consisting of binary association pairs, for instance PPI's, protein-DNA/RNA, or protein-metabolite interactions. **Figure 2** shows the essential biological molecular interactions governing cell behaviour in an over-simplified

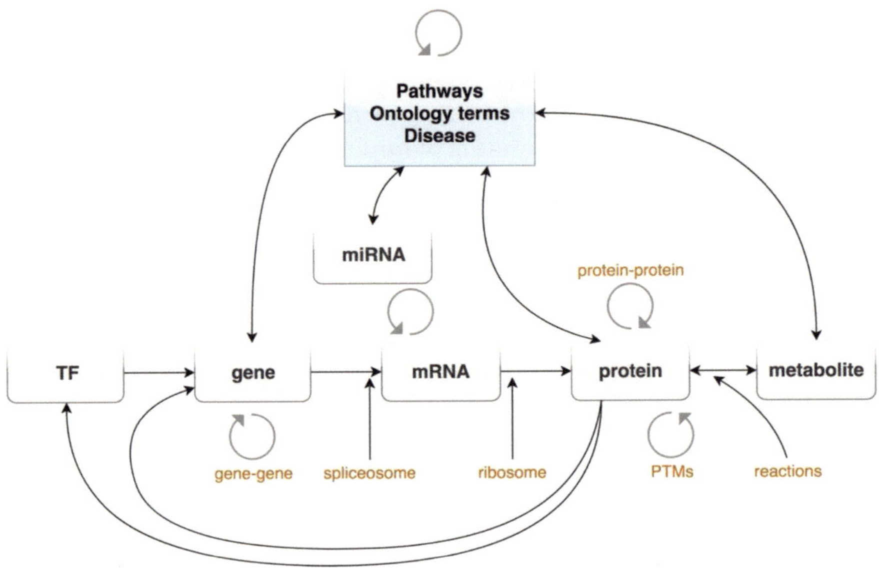

Figure 2.
Description of the essential known relationships/interactions in an over-simplified biological system. Transcription factor (TF), microRNA (miRNA), post-translational modifications (PTMs). The illustration does not account for epigenetic modifications, for instance DNA methylation and histone modifications known to occur and regulate gene expression. Dark coloured arrows denote entity associations, while self-circular arrows describe self-pair interactions or modifications.

biological system. A curated compilation of high-quality sources of binary interactions is considered a prime resource in the Systems Biology field and thereby enabling a deeper understanding of the larger picture—be it at the level of the organism, organ, tissue, or cell—by putting its components together. It's in stark contrast to decades of reductionist biology, which merely focuses on the properties of its individual components [13]. Most disease conditions exhibit expression of complex disease phenotypes [13], such as obesity, metabolic syndrome, autoimmune diseases and renal diseases.

Using the words of Ronald Germain to provide a definition of Systems Biology, he advocates that: "There are an endless number of definitions, it's even worse than the elephant," that infamous elephant that stymies the attempts of blind men to describe it because each feels just one part, "Some people think of it as bioinformatics, taking an enormous amount of information and processing it." "The other school of thought thinks of it as computational biology, computing on how the systems work. You need both parts." Ironically, to best understand this novel approach, we should take a reductionist approach to defining its parts. The system, it seems, is more than the sum of its parts [14]. Systems Biology requires comprehensive data at all molecular levels, a profound understanding of biological systems, data-criteria based assessment and in-deep understanding of the limitations of the techniques used in the experimental setup. Moreover, systems biology requires prior knowledge either published or sourced from biological databases and newly predicted and frequent molecular events requires further *in vivo/vitro* validation [15]. Systems Biology is cross-disciplinary: "[...] a scientific approach that combines the principles of engineering, mathematics, physics, and computer science with extensive experimental data to develop a quantitative as well as a deep conceptual understanding of biological phenomena, permitting prediction and accurate simulation of complex (emergent) biological behaviours" (Ronald Germain in [14]). Furthermore, systems biology promotes understanding of the functional roles and interplays of all molecules in cells in health and disease. Also provides a framework for large-scale data-driven analysis and predictions based on prior knowledge of experimentally identified interactions and pathways [16]. Thus, more relevant that the underlying high-throughput screening methods, including genomics, proteomics, metabolomics, and also bioinformatics approaches is the use of such methods in a integrative manner to holistically understand how nonlinear processes and their outcomes are regulated in a biological system [17].

3.1 Bridging the gap between fields

Over the last 10 years, major efforts to reclassify diseases based on molecular insights from advances in molecular biology, bioinformatics and high-throughput screening yielded novel disease subtypes among many disease conditions. The use of multiple data types, including clinical endpoints—omics and ontology-based data have been used to reconstitute disease phenotypes, classify and to refine disease-relationships [18]. Nevertheless, the development of a molecular-based disease taxonomy that links global molecular networks with pathological phenotype landscapes remains elusive. Systems medicine can be perceived as a multi-disciplinary collaborative effort driven by the application of systems biology approaches, which includes methodological workflows from high-throughput-omics technologies to generate data, warehousing management systems for data flow and handling and methods for data analytics and interpretation in the context of biomedical research [19]. Ultimately, with further adoption of a systems-based approach patients will benefit of a measurable improvement of their health status since processes of disease onset and progression will be mechanistically identified, leading to new insights regarding disease-disease boundaries, and disease subtyping which facilitates ideal pharmacological interventions as drug repurposing [20]. For instance, the identification of digoxin, a drug used as therapy for atrial fibrillation and congestive heart failure [21] as potential drug candidate for pharmacological intervention in medulloblastoma subtypes 3 and 4 [22]. The authors of the study implemented an integrative systems biology approach using genomic data and collating existing drug-drug, drug-targets interactions information into a tridimensional functional-drug network. This approach involved handling omic data sets such as DNA-seq—mutated genes, copy-number variation (CNV)—repeated sections of the genome, RNA-seq and methylation profiles, combined with clinical measurements of patient outcomes (survival data) and fused using network-based and probabilistic methods that yielded a network composite with disrupted driver signalling networks and potential drug candidates [22].

4. Large-scale data: omics platforms

The advent of new high-throughput technologies (sequencing, array-based and mass spectrometry) led to an explosion of available data, not only by the number of experiments performed, but also by the data density obtained per experiment. Here, we will provide description of detection platforms handling molecular datasets; for medical imaging data types and analysis strategies please see the following review [23].

4.1 DNA microarrays and next-generation sequencing (NGS)

Microarray technologies have been widely used in research for primary screening, including gene expression profiling and providing genotype-phenotype relationship. Moreover, if properly designed, microarrays will not only provide information on gene expression and expressed single nucleotide polymorphisms (SNPs), but also detect exon junctions and fusion genes [24]. However, identical to PCR-based techniques, the design of probes requires prior knowledge. Therefore, microarrays are mostly applied in the quantification of known sequences and not for the discovery of new variants, transcripts or other unknown features [25]. Microarrays have numerous limitations. For instance, they render an indirect measurement of the relative concentration of a particular nucleic acid sequence [26]. Another limitation is based that a DNA-array can only detect sequences that the array was designed for. In addition, non-coding RNA's that are not yet recognised as expressed are typically not represented on an array [26]. Microarrays are still considered a reliable technique for routine and/or initial screening that allows multiplex quantitation of microRNAs and gene probes expression in a fast, simple and affordable way. Nevertheless, the continuous drop in the cost of NGS at a level that virtually matches the cost of DNA microarray-based platforms, thus is foreseen that DNA-arrays will be fully replaced by sequencing methods within the next decade [26].

4.2 Proteomics

The use of omics technologies, including quantitative proteomics methods aims to identify and quantify the dynamics of protein abundance, in order to gain a deeper understanding of the associated biological functions. Thereby, the quantification of the expression level and state of all proteins at a given time can characterise physiological-states at the cellular-level [27]. Mass spectrometry (MS) technology, particularly tandem mass spectrometry (MS/MS), has been utilised as a discovery engine in proteomics [28]. This technology allows for identification and simultaneously quantification of hundreds or even thousands of proteins in an experimental setup, which enables real-time comparisons for instance between two or more physiological states [29]. Furthermore, peptide sequence composition will directly impact on ionisation efficiency, and their intensities observed in a spectrum often do not reflect their abundances, [30] thereby many label-free or label-based quantitation methods have arisen to allow comparative proteomic analysis. For instance, label-free proteomic approaches such as ion intensity, spectral counting have a simplified workflow when compared to labelling techniques; have no theoretical limit concerning multiplexing capability providing an improved proteome coverage, but lower quantification accuracy when compared with labelling methods (e.g. iTRAQ: isobaric tags for relative and absolute quantitation, SILAC: stable isotope labelling by/with amino acids in cell culture) [30]. In proteomics, several algorithms have been developed to query and cross compare MS data. The most popular used to identify proteins from raw MS data are for instance, MASCOT, SeQuest, OMSSA, X!Tandem [31], Andromeda [32], MS-GF [33], Paragon [34] and more recently, Morpheus [35] and an improved SEQUEST-like algorithm—ProLuCID [36]. The rise in the number of algorithms and specialised computational tools for analysis of MS-based proteomics data sets led to the development of workflows/pipelines such as PEAKS [37], MaxQuant [38], OpenMS Proteomics Pipeline (TOPP) [39], Trans-Proteomic Pipeline (TPP) [40] and others for further downstream data analysis—Perseus [41].

4.3 Metabolomics

In many metabolomics studies the identification and quantification of metabolites mainly rely on the application of analytical methods based on mass spectrometry (MS) (either coupled with a liquid or gas-chromatograph) and nuclear magnetic resonance (NMR) spectroscopy [42]. Metabolites are defined as small molecules, usually less than 1000 Da, which suffer several changes during cellular metabolism [43]. The selection of a particular platform depends upon the aims of the experimental study and is typically driven by establishing a compromise among sensitivity, specificity, and scanning speed [44]. Metabolomics approaches can be globally split either by the full range measurement/analysis of all compounds in a given sample—untargeted metabolomics, or targeted metabolomics, in which a set of predefined and biochemically well-characterised compounds are measured in a sample [44]. MS has become an essential method for non-targeted profiling of metabolites in complex bio-samples, particularly low-abundance metabolites, due to its high sensitivity and selectivity capabilities when using liquid chromatography (LC) coupled to tandem MS/MS [45]. Metabolomics data from NMR and MS platforms are complex because they usually contain thousands distinct peaks therefore, multivariate statistical analysis plays an important role in metabolomics for reducing data dimensions, differentiating similar spectra, and in the development of predictive models [46]. Metabolomics is used as a screening tool in current healthcare settings, and could be greatly utilised to monitor therapy efficacy, and assess potential drug side-effects [47].

5. Data-driven approaches and multi-omics data integration

In the field of biomedical research adopting an unbiased approach or "hypothesis-free" (depending of the author and field of study, also defined as hypothesis-generating approach, data-driven research, or discovery research) to research can bring several benefits when compared with the widely used scientific approach—hypothesis-driven research (traditional approach). In which, the latter, in some cases encourages poor scientific practices by forcing/imposing qualitative and weak hypotheses that Are not prepared for strong statistical inference or quantitative analysis (QA) modelling, thereby in such cases an explicitly exploratory approach should be set as default [48]. In order to overcome this problem, large-scale approaches such as expression profiling started to become very popular in the mid '90s, and beginning of 2000, with the advent, rapid development and availability of high-throughput mass spectrometry, other methods followed [49]. Computational methods to analyse this flood of data were developed accordingly, however the majority only focused on one specific technology or experimental setup and up to this day are very often not interchangeable in other technological platforms. Large-scale approaches employed in omics research need a different analysis methodology, which is especially true if integrative analysis techniques are employed. True integrative (as opposed to integrating linear relationship data such as gene-protein data) approaches go beyond simple data fusion and gave rise to the field of Systems Biology. On the other hand, hypothesis-generating research (systems biology-derived hypotheses) and hypothesis-driven research are complementary, thus combining both approaches will certainly sustain more chances of a complete understanding of complex biological systems, than either approach on its own [48]. With the advent of high-throughput technologies their application in the biomedical field was a foreseen logical step. However, until recently integration of multi-omic data was not a common approach in former analysis workflows. The literature and publicly available databases are awash with data, yet the main approach of integrating all this information in a disease-specific context is traditionally based on meta-analysis at best or cannot be accomplished using standard computational methods. This molecular information can then be integrated in a further stage by means of meta-analysis or by cross-normalisation of data from different acquisition platforms [50]. A combinatorial stepwise data integration (**Figure 3**) approach can be used in order to incorporate data from different biological layers of information to predict phenotypic outcomes [51]. On the other side, by recreating the cell

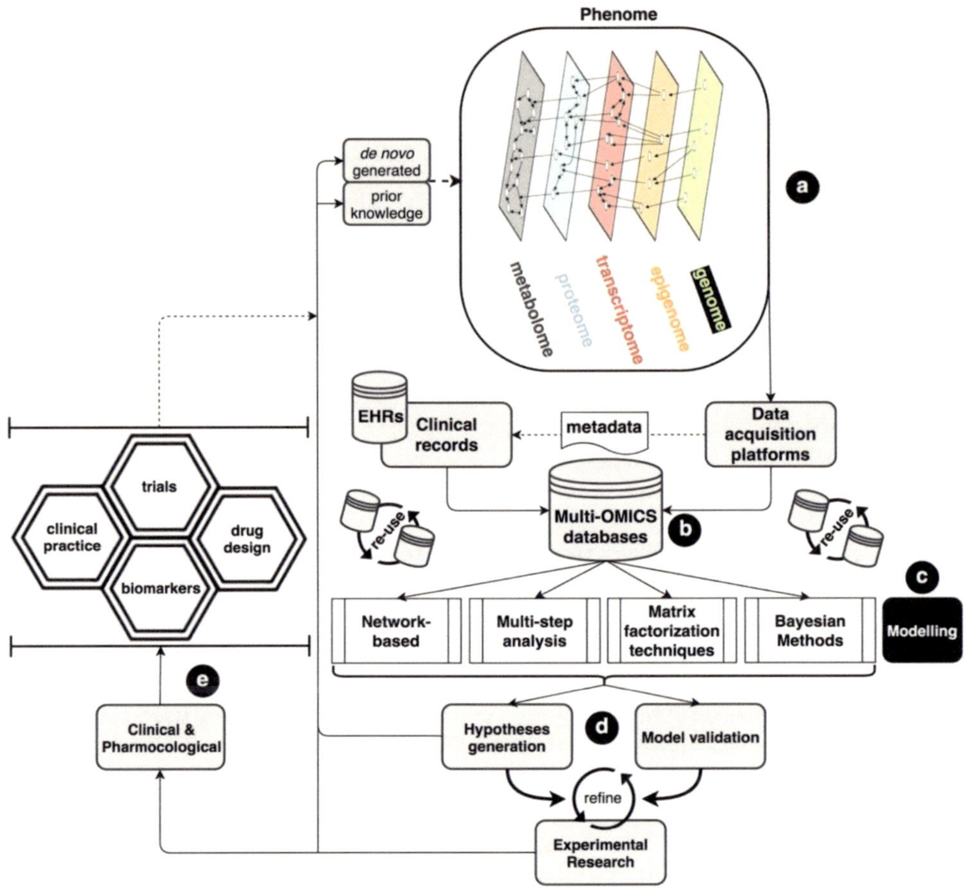

Figure 3.
Purposed workflow for a data-driven approach. Data generation from omics platforms plus existing biological information (a), development of a multi-omics database (b), selection of suitable modelling methods (c), model validation and use for hypothesis-generating research (d), lead optimization and candidate selection (e).

environment and dynamics by describing their interactions on a qualitative and quantitative manner and relying on underlying data (prior biological knowledge) for connectivity, e.g. PPI's, molecular co-occurrence, ontologies and enzymatic reactions [52]. Large-scale data sets for instance derived from multi-omics platforms may also be used to infer novel relationships by network learning approaches using Bayesian inference models [51] and extracting molecular information from multi-layered networks. This approach (as in many others) is challenging since it requires enough statistical power, higher number of samples to deduce all the possible interactions. Another challenge is due to the lack of uniformisation regarding the 'gold 'standards (criterions for evaluation) for accepting or rejecting relationships of the inferred model; however the ability to recreate a well-accepted interaction can at least be used for benchmarking methods in biological systems [53].

6. Biological databases and database systems

Databases form the basis for most applications in bioinformatics. The number of biological databases available now is enormous, the journal of Nucleic Acids Research (NAR) catalogues a total of 1737 molecular biology databases (2018 edition) [54]. The 2018 edition contains an enormous set of 181 papers that describe the adding of 82 new biological databases, 84 updates and as well 15 databases published elsewhere. However, a prominent issue concerns that many databases are not maintained over time and abandoned, yet they persist in database listings. There are many different types of databases, ranging from primary databases containing sequence data such as nucleic acid or protein; secondary databases or also known as pattern databases hosts, that results from the analysis of the sequences held in primary databases.

6.1 General scope expression databases

The Gene Expression Omnibus (GEO) [55] is a public repository that functions as both warehouse of raw microarray and other gene-based high-throughput data, and additionally serves as a platform for gene differential expression (DE) analysis using the GEO2R tool across a multitude of experimental conditions of user-submitted pre-processed data sets. In the same way, the European counterpart for storing of high-throughput genomics exists such as the European Bioinformatics Institute (EMBL-EBI) throughout the ArrayExpress database [56]. These data resources are both in compliance with community guidelines for description of an experimental setup for microarray and high-throughput NGS experiment. Comparatively, there is currently much less support for sharing of proteomics and metabolomics data sets despite the increasing demand. Public efforts for proteomic data sharing yielded the Proteomics Identification Database (PRIDE) that contains over 10,100 user-submitted MS-based raw proteomic data sets (September 2018) [57]. PeptideAtlas [58] handles re-analysed data sets via the TPP pipeline to provide end-users a consistently view over their data. MetaboLights [59] hosts user-submitted metabolomics experiments, which currently houses 439 experiments (November 2018). The standards for reporting proteomics and metabolomics experiments are coordinated by the Human Proteome Organisation's Proteomics Standards Initiative (HUPO-PSI), and Metabolomics Standards Initiative (MSI) respectively.

6.2 Disease profiling databases

Our group developed more specialised databases resources in several disease conditions handling pre-selected data sets containing DE molecules. In nephrology, we developed the Chronic Kidney Disease database (CKDdb) [60] storing microRNA, genomics, peptidomics, proteomics and metabolomics information relevant to CKD, collected from over 300 studies in the literature and integrated into the Pan-omics Analysis DataBase (PADB). The PADB framework (www.padb.org) uses gene and protein clusters (CluSO) and mapping of orthologous genes (OMAP) between species therefore facilitating data harmonisation from a diverse range of omics platforms and across several species, which makes it an invaluable resource for systems biology data-driven approaches. Also, many conditions associated with the cardiovascular system are covered in the Cardio/Vascular Disease (C/VD) database [61], which gives special emphasis on coronary artery disease (CAD). In neurological associated conditions such as Multiple Sclerosis we also developed the MuScle database [62] that stores and integrates curated data sets mined from large-scale studies with focus on genomics and miRNA. Likewise, we built a cancer-related differential expression database: the Multi-Omics Cancer database (MoCadb) that integrates clustered molecular information covering multi-omics studies in many gastro-intestinal cancers. In the same framework we also cover an assorted disease profiling database valuable for subtractive disease analysis studies, the Large-Scale Screening Resource (LSSR) that contains 81,980 entries, referring to 13,589 molecules. Moreover, a peak profiling database for biomarker patterns research, the Urinary Peptidomics and Peak-maps (UPdb) [63] database that comprises Human urinary fingerprints from 200 subjects analysed mainly through surface enhanced laser desorption ionisation-time of flight mass spectrometry (SELDI-TOF-MS).

7. Software tools and solutions

Many modern high-throughput technologies lead to the generation of exceptionally large-scale and complex datasets, which includes PPI's, protein-DNA interactions, kinase-substrate interactions, qualitative and quantitative genetic-interactions gene co-expression [64]. The "Big Data" challenge can be fulfilled by the development of Bioinformatics tools to handle these large-datasets to reduce their complexity to a level that enables rationale interpretation and in this way is more likely to provide new biological insights to the Life Sciences. The compilation (not an exhaustive list) of many web-based, standalone tools and R-based packages are described in **Table 1**. They allow the accomplishment of different-omics tasks

Name	Description	Webpage	Ref.
iClusterPlus	Integrative clustering	bioconductor.org/packages/iClusterPlus	[84]
mixomics	Data integration (CCA,PLS,PCA)	mixomics.org	[85]
omicade4	MCIA and CIA	bioconductor.org/packages/omicade4	[86]
pwOmics	Pathway-based integration of omics	bioconductor.org/packages/pwOmics	[87]
PRESTO	Dimensionality reduction of multivariate data	github.com/saramcardle/PRESTO	[88]
caret	Classification and regression training	cran.r-project.org/web/packages/caret	—
GEO2R	Identify DE genes using GEOquery & limma R packages	ncbi.nlm.nih.gov/geo/geo2r	[55]
Metabo Analyst	Metabolomics analysis	metaboanalyst.ca	[89]
Networkanalyst/INMEX	Integration of gene DE via network approaches	networkanalyst.ca	[90]
ExAtlas	Meta-analysis & visualisation of gene DE	lgsun.irp.nia.nih.gov/exatlas	[91]
Elastic net	Gene DE with fitted GLM	https://zenodo.org/record/16006	[92]
ATHENA	Integration of genomics with clinical data	ritchielab.org/software/athena-downloads	[93]
Network propagation	Gene DE, mutations, PPI's	http://apps.cytoscape.org/apps/Diffusion	[94]

PMA, Penalised Multivariate Analysis; RGCCA, Regularised and Sparse Generalised Canonical Correlation Analysis for Multiblock Data; caret, Classification and REgression Training; ATHENA, Analysis Tool for Heritable and Environmental Network Associations; CCA, Canonical-Correlation Analysis; PLS, Partial Least Squares; PCA, Principal Component Analysis; CIA, Co-Inertia Analysis; MCIA, Multiple Co-Inertia Analysis; GO, gene ontology; DE, differential expression; GLM, generalised linear models.

Table 1.
Web-based, standalone tools and R packages dedicated to different-omics tasks such as feature selection, sample classification, multivariate approaches in data integration and meta-analysis.

such as feature selection, sample classification, multivariate methods. Cytoscape [65] is a tool primarily designed for network visualisation and analysis and has useful plugins available through the hosting website. Cytoscape makes use of a wide wealth variety of plugins to extend its functionality which are designed by the scientific community. The platform counts with several freely available apps/plugins (over 300 apps available on November 2018) for a diverse array of uses and analysis types.

7.1 Gene ontology (GO) and pathway-term-enrichment

The Gene Ontology (GO) consortium [66] aims to capture the increasing knowledge on gene function in a controlled vocabulary applicable to a wide range of organisms. GO represents genes and gene products attributes on matters of their associated biological processes (BP), cellular components (CC) and molecular functions (MF). GO is considered roughly hierarchical, with 'child' elements (terms) being more specific than their 'parent' elements (terms), nevertheless, a 'child' element (term) might have more than one parent element. The ClueGO app [67] is used for the integration and visualisation of GO and pathway terms sourced from KEGG [68], WikiPathways [69] and Reactome [70]. The resultant ClueGO network is established based in kappa statistics which shows the agreement on how any given gene and/or gene products pairs share similar terms. The ClueGO analysis output is conditioned by thresholding of the kappa coefficient, in which a higher coefficient conducts only to the visualisation of close-related terms with very identical gene products. While, lower kappa coefficients will let visualisation of less associated terms.

7.2 Gene-disease associations (GDA)

The conclusion of the Human Genome Project led to the massification of research related with uncovering genotype—disease phenotype associations [71]. This event translated in a disparate growth in the number of publications and on the other side a limited and slow paced biocuration of these newly discovered evidences. Currently, DisGeNET [72] unifies biomedical literature evidence based on GDA collated from a multitude of databases. This database makes use of the Medical Subjects Headings (MeSH) tree structure for disease classification by a Unified Medical Language System. The potential of the database is extended by disgenet2r package and optional programmatic access.

7.3 Protein-protein interactions (PPIs)

STRING database [6] collates molecular information to cover both known and predicted PPI's. All molecular interaction data is originally from primary interaction databases such as IntAct [73], BioGRID [74] and additional text-mining, coexpression and high-throughput experiments and computationally predicted PPIs. The up-to-date database version 10.5 comprises nearly 26 million PPI with a confidence score greater than 0.9 of more than 9 million proteins across 2031 organisms. GeneMANIA is another source for PPIs analysis and is accessible via web interface [75], and also as a Cytoscape app that can be used to detect related genes of a input query by means of a "guilt-by-association" strategy, which explores the realisation that a protein function can be obtained from another by seeing whether it interacts with another of known function. The app uses a large database of functional interaction networks, indexing 2152 association networks containing more than 500 million interactions mapped to 166,084 genes from nine organisms.

7.4 Combining metabolomic and gene expression data

Multi-omics datasets might not only contain protein and gene data, but also expression profiles of chemical compounds. While it is easy and straightforward to combine protein/DNA/RNA expression data using common identifiers, this is not the case for metabolism end-products—metabolites. This requires a guilt-by-association, which explores the rationale that metabolites are frequently produced by enzymes and a shift in metabolite expression can reflect an up-stream shift in protein or gene expression. This involves semantic searches in enzyme repositories—BRENDA to identify potential proteins and has some inherent pitfalls such as uncertainty which enzyme/isoform is responsible for the metabolic change. Additionally, the same compound could also be generated by several proteins, which adds to the uncertainty. Therefore, metabolic datasets are often treated as separate entities in multi-omics studies and analysed independently and then converged only at the level of final outcomes [76]. The MetScape 3 app [77] for the Cytoscape can perform joint analysis of both metabolomic and gene expression data and allows visualisation of the entire fused network, or by selecting custom views based on metabolic pathways When dealing with large-scale datasets, there is the option to use a concept file based on pre-computed gene set enrichment analysis (GSEA), along with statistical and fold-change thresholds.

7.5 Transcription factor (TF)-driven modules and microRNA-target regulation

Transcription factors (TF) are critical for the regulation of gene expression since they control if gene's DNA is transcribed into RNA [78]. A compendium on non-redundant TF and TF binding sites can be found at JASPAR [79]. The number of human TF ranges from 1500 to 2600, depending on source and stringency [78]. Direct analysis of modulated events due to TFs is not only valuable but might shed light on hidden elements that conventional pathway analysis cannot reveal. However, many TF binding sites and modulated genes are very hypothetical and often a random guess. Therefore, network-based analysis and interpretation involving TF elements should be taken with caution. CyTargetLinker [80] for extends existing biological networks by adding interactions associated with regulatory elements such as TF-target, miRNA-target or drug-targets. The application requires a loaded network with network attributes preferentially mapped to Ensembl, NCBI gene, UniProt, miRBase or DrugBank. Similarly, in CluePedia [81] users can perform miRNA analysis, by matching it to target-genes via selection of different database resources custom versions. Users can upload a list of genes and query the app to perform gene/miRNA enrichments. Then it will generate a miRNA-target interaction network that can be reused for inline integration with GO and pathway term clustering [81] within ClueGO.

7.6 Pathway mapping and visualisation

7.6.1 PathVisio pathway mapping and edition

PathVisio [82] allows drawing, edition, and visualisation of pathways handling gene, protein and metabolite data that can be further cross-mapped via the BridgeDb [83]. Inference of relevant pathways is based on an archive of pre-existent pathway maps from WikiPathways [69] and Reactome [70], establishing pathway over-representation based on a Z-score statistical procedure under the hypergeometric distribution and a *P*-value ranking based on a permutation procedure (randomisation test) that compares actual and permuted Z-scores. Pathways with a permuted $P < 0.05$ are considered significant by default.

7.6.2 KEGG pathway mapping

KEGG is an integrated database resource of biological systems integrating genomic, compound and functional information. KEGG allows analysis of datasets from high-throughput omics technologies by uploading a list of genes/proteins or metabolites along with optional statistical scores and fold-change values. After converting to KEGG internal identifiers, the molecular data is matched (KEGG mapper) into a collection of curated pathways, covering metabolism, signalling transduction pathways, specific pathways for several disease conditions and drug development.

8. Conclusions and future perspectives

The availability of large-scale multi-omics data has opened the avenue to gain an unrivalled insight in disease-associated molecular pathophysiological changes. Simultaneously it has become apparent that systems to integrate and correlate this data are either inadequate or non-existent. The literature and publicly available databases are awash with data, yet the main approach of integrating all this information in a disease-specific context is traditionally based on meta-analysis at best or cannot be accomplished using standard computational methods. In order to better model complex organisms, samples from multiple tissues of the same individuals should be studied simultaneously using omics data, which will require the development of novel analysis methods. Acquiring the relevant tissues and/or body fluid sources from Human study cohorts can of course be difficult, thereby comparative systems biology may help identify which organisms may be similar enough in each aspect to be used as models. It is sometimes suggested that omics technologies and systems biology have failed to deliver many breakthrough enhancements to the treatment of complex diseases. In some cases, it may be that in fact such diseases are not truly one disease from a system or reductionist point-of-view, but several with the same or similar phenotypic end-points—i.e., with the current terminology they are unknown subtypes of disease. If this is the case, then the overlap between the systems is poor and statistical methods which the approach relies on require very large cohorts for identification of these subtypes and subsequent description of each system. Other possibilities are that longitudinal data or samples from different tissues are required. Other relevant concerns arise from biomarker validation studies, such as correlated observations (i.e. multiple observations per patient), multiplicity (testing multiple biomarkers or endpoints), multiple clinical endpoints (interest in more than one relevant endpoint) and selection bias (from retrospective data or observational study). Data-driven investigations using systems biology approaches, although offer complete views over the function of biological systems in health and disease its limited by the state of completeness of prior biological information.

Acknowledgements

The research leading to these results has received funding from the European Union's Seventh Framework Programme FP7/2007–2013 under grant agreement FP7-PEOPLE-2013-ITN-608332. The funder had no role in study design, data collection and analysis, decision to publish, or preparation of the manuscript.

Conflict of interest

The authors declare that there is no conflict of interest regarding the publication of this manuscript.

Author details

Marco Fernandes[1] and Holger Husi[1,2*]

1 Institute of Cardiovascular and Medical Sciences, BHF Glasgow Cardiovascular Research Centre, University of Glasgow, Glasgow, United Kingdom

2 Division of Biomedical Sciences, Centre for Health Science, University of the Highlands and Islands, Inverness, United Kingdom

*Address all correspondence to: holger.husi@uhi.ac.uk

References

[1] World Health Organization. ICD-10: International Statistical Classification of Diseases and Related Health Problems: tenth revision. Geneva: World Health Organization. 2004

[2] National Research Council Committee on A Framework for Developing a New Taxonomy of Disease. Toward Precision Medicine. Building a Knowledge Network for Biomedical Research and a New Taxonomy of Disease. Washington, DC: National Academies Press; 2011

[3] Zhou X, Lei L, Liu J, Halu A, Zhang Y, Li B, et al. A Systems Approach to Refine Disease Taxonomy by Integrating Phenotypic and Molecular Networks. eBioMedicine. 2018;**31**:79-91

[4] Hu JX, Thomas CE, Brunak S. Network biology concepts in complex disease comorbidities. Nature reviews Genetics. 2016;**17**(10):615-629

[5] Jensen AB, Moseley PL, Oprea TI, Ellesoe SG, Eriksson R, Schmock H, et al. Temporal disease trajectories condensed from population-wide registry data covering 6.2 million patients. Nature Communications. 2014;**5**:4022

[6] Szklarczyk D, Franceschini A, Wyder S, Forslund K, Heller D, Huerta-Cepas J, et al. STRING v10: Protein-protein interaction networks, integrated over the tree of life. Nucleic Acids Research. 2015;**43**(Database issue):D447-D452

[7] Bailey P, Chang DK, Nones K, Johns AL, Patch AM, Gingras MC, et al. Genomic analyses identify molecular subtypes of pancreatic cancer. Nature. 2016;**531**(7592):47-52

[8] Lee SY, Lee SH, Yang EJ, Kim EK, Kim JK, Shin DY, et al. Metformin ameliorates inflammatory bowel disease by suppression of the STAT3 signaling pathway and regulation of the between Th17/Treg balance. PLoS One. 2015;**10**(9):e0135858

[9] Reeve J, Bohmig GA, Eskandary F, Einecke G, Lefaucheur C, Loupy A, et al. Assessing rejection-related disease in kidney transplant biopsies based on archetypal analysis of molecular phenotypes. JCI Insight. 2017;**2**(12):pii:94197

[10] Parkes M, Reeve J, Kim D, Macdonald P, Aliabadi A, Goekler J, et al. Molecular assessment of heart transplant biopsies: Emergence of the injury dimension. Transplantation. 2018;**102**:S62-SS3

[11] Birner P, Prager G, Streubel B. Molecular pathology of cancer: How to communicate with disease. ESMO Open. 2016;**1**(5):e000085

[12] Rolland T, Tasan M, Charloteaux B, Pevzner SJ, Zhong Q, Sahni N, et al. A proteome-scale map of the human interactome network. Cell. 2014;**159**(5):1212-1226

[13] Chuang HY, Hofree M, Ideker T. A decade of systems biology. Annual Review of Cell and Developmental Biology. 2010;**26**:721-744

[14] Germain RN. Systems-biology-as-defined-by-nih. Available from: https://irp.nih.gov/catalyst/v19i6/systems-biology-as-defined-by-nih

[15] Ashburn TT, Thor KB. Drug repositioning: Identifying and developing new uses for existing drugs. Nature Reviews Drug Discovery. 2004;**3**(8):673-683

[16] Robinson SW, Fernandes M, Husi H. Current advances in systems and integrative biology. Computational and Structural Biotechnology Journal. 2014;**11**(18):35-46

[17] Gottschalk RA, Martins AJ, Sjoelund V, Angermann BR, Lin B, Germain RN. Recent progress using systems biology approaches to better understand molecular mechanisms of immunity. Seminars in Immunology. 2013;**25**(3):201-208

[18] Menche J, Sharma A, Kitsak M, Ghiassian SD, Vidal M, Loscalzo J, et al. Disease networks. Uncovering disease-disease relationships through the incomplete interactome. Science (New York, N.Y.). 2015;**347**(6224):1257601

[19] Apweiler R, Beissbarth T, Berthold MR, Bluthgen N, Burmeister Y, Dammann O, et al. Whither systems medicine? Experimental & Molecular Medicine. 2018;**50**(3):e453

[20] Ayers D, Day PJ. Systems medicine: The application of systems biology approaches for modern medical research and drug development. Molecular Biology International. 2015;**2015**:698169

[21] Virgadamo S, Charnigo R, Darrat Y, Morales G, Elayi CS. Digoxin: A systematic review in atrial fibrillation, congestive heart failure and post myocardial infarction. World Journal of Cardiology. 2015;**7**(11):808-816

[22] Huang L, Garrett Injac S, Cui K, Braun F, Lin Q, Du Y, et al. Systems biology-based drug repositioning identifies digoxin as a potential therapy for groups 3 and 4 medulloblastoma. Science Translational Medicine. 2018;**10**(464):pii:eaat0150

[23] Papp L, Spielvogel CP, Rausch I, Hacker M, Beyer T. Personalizing medicine through hybrid imaging and medical big data analysis. Frontiers in Physics. 2018;**6**:51

[24] Mantione KJ, Kream RM, Kuzelova H, Ptacek R, Raboch J, Samuel JM, et al. Comparing bioinformatic gene expression profiling methods: microarray and RNA-Seq. Medical Science Monitor Basic Research. 2014;**20**:138-142

[25] Mortazavi A, Williams BA, McCue K, Schaeffer L, Wold B. Mapping and quantifying mammalian transcriptomes by RNA-Seq. Nature Methods. 2008;**5**(7):621-628

[26] Bumgarner R. Overview of DNA microarrays: Types, applications, and their future. Current Protocols in Molecular Biology. 2013; Chapter 22, Unit 22.1:1-11

[27] Cox J, Mann M. Is proteomics the new genomics? Cell. 2007;**130**(3):395-398

[28] Yates JR, Ruse CI, Nakorchevsky A. Proteomics by mass spectrometry: approaches, advances, and applications. Annual Review of Biomedical Engineering. 2009;**11**:49-79

[29] Vizcaino JA, Cote RG, Csordas A, Dianes JA, Fabregat A, Foster JM, et al. The PRoteomics IDEntifications (PRIDE) database and associated tools: Status in 2013. Nucleic Acids Research. 2013;**41**(Database issue):D1063-D1069

[30] Lindemann C, Thomanek N, Hundt F, Lerari T, Meyer HE, Wolters D, et al. Strategies in relative and absolute quantitative mass spectrometry based proteomics. Biological Chemistry. 2017;**398**(5-6):687-699

[31] Keshava Prasad TS, Goel R, Kandasamy K, Keerthikumar S, Kumar S, Mathivanan S, et al. Human protein reference database—2009 update. Nucleic Acids Research. 2009;**37**(Database issue):D767-D772

[32] Cox J, Neuhauser N, Michalski A, Scheltema RA, Olsen JV, Mann M. Andromeda: A peptide search engine integrated into the MaxQuant environment. Journal of Proteome Research. 2011;**10**(4):1794-1805

[33] Kim S, Gupta N, Pevzner PA. Spectral probabilities and generating functions of tandem mass spectra: A strike against decoy databases. Journal of Proteome Research. 2008;**7**(8):3354-3363

[34] Shilov IV, Seymour SL, Patel AA, Loboda A, Tang WH, Keating SP, et al. The Paragon Algorithm, a next generation search engine that uses sequence temperature values and feature probabilities to identify peptides from tandem mass spectra. Molecular & Cellular Proteomics. 2007;**6**(9):1638-1655

[35] Wenger CD, Coon JJ. A proteomics search algorithm specifically designed for high-resolution tandem mass spectra. Journal of Proteome Research. 2013;**12**(3):1377-1386

[36] Xu T, Park SK, Venable JD, Wohlschlegel JA, Diedrich JK, Cociorva D, et al. ProLuCID: An improved SEQUEST-like algorithm with enhanced sensitivity and specificity. Journal of Proteomics. 2015;**129**:16-24

[37] Ma B, Zhang K, Hendrie C, Liang C, Li M, Doherty-Kirby A, et al. PEAKS: Powerful software for peptide de novo sequencing by tandem mass spectrometry. Rapid Communications in Mass Spectrometry. 2003;**17**(20):2337-2342

[38] Cox J, Matic I, Hilger M, Nagaraj N, Selbach M, Olsen JV, et al. A practical guide to the MaxQuant computational platform for SILAC-based quantitative proteomics. Nature Protocols. 2009;**4**(5):698-705

[39] Kohlbacher O, Reinert K, Gropl C, Lange E, Pfeifer N, Schulz-Trieglaff O, et al. TOPP--the OpenMS proteomics pipeline. Bioinformatics. 2007;**23**(2):e191-e197

[40] Deutsch EW, Mendoza L, Shteynberg D, Farrah T, Lam H, Tasman N, et al. A guided tour of the trans-proteomic pipeline. Proteomics. 2010;**10**(6):1150-1159

[41] Tyanova S, Temu T, Sinitcyn P, Carlson A, Hein MY, Geiger T, et al. The Perseus computational platform for comprehensive analysis of (prote)omics data. Nature Methods. 2016;**13**(9):731-740

[42] Dunn WB, Broadhurst DI, Atherton HJ, Goodacre R, Griffin JL. Systems level studies of mammalian metabolomes: The roles of mass spectrometry and nuclear magnetic resonance spectroscopy. Chemical Society Reviews. 2011;**40**(1):387-426

[43] Patti GJ, Yanes O, Siuzdak G. Innovation: Metabolomics: The apogee of the omics trilogy. Nature Reviews Molecular Cell Biology. 2012;**13**(4):263-269

[44] Lei Z, Huhman DV, Sumner LW. Mass spectrometry strategies in metabolomics. The Journal of Biological Chemistry. 2011;**286**(29):25435-25442

[45] Dettmer K, Aronov PA, Hammock BD. Mass spectrometry-based metabolomics. Mass Spectrometry Reviews. 2007;**26**(1):51-78

[46] Blekherman G, Laubenbacher R, Cortes DF, Mendes P, Torti FM, Akman S, et al. Bioinformatics tools for cancer metabolomics. Metabolomics: Official journal of the Metabolomic Society. 2011;**7**(3):329-343

[47] Trivedi DK, Hollywood KA, Goodacre R. Metabolomics for the masses: The future of metabolomics in a personalized world. New Horizons in Translational Medicine. 2017;**3**(6):294-305

[48] Carroll S. Defining the scientific method: Editorial. Nature Methods. 2009;**6**(4):237

[49] Zhang Z, Wu S, Stenoien DL, Pasa-Tolic L. High-throughput proteomics.

Annual Review of Analytical Chemistry (Palo Alto, CA). 2014;7:427-454

[50] Walsh CJ, Hu P, Batt J, Santos CC. Microarray meta-analysis and cross-platform normalization: Integrative genomics for robust biomarker discovery. Microarrays (Basel). 2015;**4**(3):389-406

[51] Huang S, Chaudhary K, Garmire LX. More is better: Recent progress in multi-omics data integration methods. Frontiers in Genetics. 2017;**8**:84

[52] Gligorijevic V, Przulj N. Methods for biological data integration: Perspectives and challenges. Journal of The Royal Society Interface. 2015;**12**(112):pii:20150571

[53] Marbach D, Costello JC, Kuffner R, Vega NM, Prill RJ, Camacho DM, et al. Wisdom of crowds for robust gene network inference. Nature Methods. 2012;**9**(8):796-804

[54] Rigden DJ, Fernandez XM. The 2018 Nucleic Acids Research database issue and the online molecular biology database collection. Nucleic Acids Research. 2018;**46**(D1):D1-d7

[55] Barrett T, Wilhite SE, Ledoux P, Evangelista C, Kim IF, Tomashevsky M, et al. NCBI GEO: Archive for functional genomics data sets--update. Nucleic Acids Research. 2013;**41**(Database issue):D991-D995

[56] Kolesnikov N, Hastings E, Keays M, Melnichuk O, Tang YA, Williams E, et al. ArrayExpress update—Simplifying data submissions. Nucleic Acids Research. 2015;**43**(Database issue):D1113-D1116

[57] Perez-Riverol Y, Csordas A, Bai J, Bernal-Llinares M, Hewapathirana S, Kundu DJ, et al. The PRIDE database and related tools and resources in 2019: Improving support for quantification data. Nucleic Acids Research. 2018;**47**:D442-D450

[58] Deutsch EW. The PeptideAtlas project. Methods in Molecular Biology. 2010;**604**:285-296

[59] Hastings J, de Matos P, Dekker A, Ennis M, Harsha B, Kale N, et al. The ChEBI reference database and ontology for biologically relevant chemistry: Enhancements for 2013. Nucleic Acids Research. 2013;**41**(Database issue):D456-D463

[60] Fernandes M, Husi H. Establishment of a integrative multi-omics expression database CKDdb in the context of chronic kidney disease (CKD). Scientific Reports. 2017;7:40367

[61] Fernandes M, Patel A, Husi HC. VDdb: A multi-omics expression profiling database for a knowledge-driven approach in cardiovascular disease (CVD). PLoS One. 2018;**13**(11):e0207371

[62] Cervantes-Gracia K, Husi H. Integrative analysis of multiple sclerosis using a systems biology approach. Scientific Reports. 2018;**8**(1):5633

[63] Husi H, Barr JB, Skipworth RJ, Stephens NA, Greig CA, Wackerhage H, et al. The human urinary proteome fingerprint database UPdb. International Journal of Proteomics. 2013;**2013**:760208

[64] Mohr S, Bakal C, Perrimon N. Genomic screening with RNAi: Results and challenges. Annual Review of Biochemistry. 2010;**79**:37-64

[65] Shannon P, Markiel A, Ozier O, Baliga NS, Wang JT, Ramage D, et al. Cytoscape: A software environment for integrated models of biomolecular interaction networks. Genome Research. 2003;**13**(11):2498-2504

[66] Ashburner M, Ball CA, Blake JA, Botstein D, Butler H, Cherry JM, et al. Gene ontology: Tool for the unification of biology. The Gene

Ontology Consortium. Nature Genetics. 2000;**25**(1):25-29

[67] Bindea G, Mlecnik B, Hackl H, Charoentong P, Tosolini M, Kirilovsky A, et al. ClueGO: A cytoscape plug-in to decipher functionally grouped gene ontology and pathway annotation networks. Bioinformatics. 2009;**25**(8):1091-1093

[68] Kanehisa M, Goto S. KEGG: Kyoto encyclopedia of genes and genomes. Nucleic Acids Research. 2000;**28**(1):27-30

[69] Kutmon M, Riutta A, Nunes N, Hanspers K, Willighagen EL, Bohler A, et al. WikiPathways: Capturing the full diversity of pathway knowledge. Nucleic Acids Research. 2016;**44**(D1):D488-D494

[70] Fabregat A, Sidiropoulos K, Garapati P, Gillespie M, Hausmann K, Haw R, et al. The reactome pathway knowledgebase. Nucleic Acids Research. 2016;**44**(D1):D481-D487

[71] Ozgur A, Vu T, Erkan G, Radev DR. Identifying gene-disease associations using centrality on a literature mined gene-interaction network. Bioinformatics. 2008;**24**(13):i277-i285

[72] Pinero J, Queralt-Rosinach N, Bravo A, Deu-Pons J, Bauer-Mehren A, Baron M, et al. DisGeNET: A discovery platform for the dynamical exploration of human diseases and their genes. Database: The Journal of Biological Databases and Curation. 2015;**2015**:bav028

[73] Orchard S, Ammari M, Aranda B, Breuza L, Briganti L, Broackes-Carter F, et al. The MIntAct project—IntAct as a common curation platform for 11 molecular interaction databases. Nucleic Acids Research. 2014;**42**(Database issue):D358-D363

[74] Stark C, Breitkreutz BJ, Reguly T, Boucher L, Breitkreutz A, Tyers M. BioGRID: A general repository for interaction datasets. Nucleic Acids Research. 2006;**34**(Database issue):D535-D539

[75] Warde-Farley D, Donaldson SL, Comes O, Zuberi K, Badrawi R, Chao P, et al. The GeneMANIA prediction server: Biological network integration for gene prioritization and predicting gene function. Nucleic Acids Research. 2010;**38**(Web Server issue):W214-W220

[76] van der Knaap JA, Verrijzer CP. Undercover: Gene control by metabolites and metabolic enzymes. Genes & Development. 2016;**30**(21):2345-2369

[77] Karnovsky A, Weymouth T, Hull T, Tarcea VG, Scardoni G, Laudanna C, et al. Metscape 2 bioinformatics tool for the analysis and visualization of metabolomics and gene expression data. Bioinformatics. 2012;**28**(3):373-380

[78] Jolma A, Yin Y, Nitta KR, Dave K, Popov A, Taipale M, et al. DNA-dependent formation of transcription factor pairs alters their binding specificity. Nature. 2015;**527**(7578):384-388

[79] Mathelier A, Fornes O, Arenillas DJ, Chen CY, Denay G, Lee J, et al. JASPAR 2016: A major expansion and update of the open-access database of transcription factor binding profiles. Nucleic Acids Research. 2016;**44**(D1):D110-D115

[80] Kutmon M, Kelder T, Mandaviya P, Evelo CT, Coort SL. CyTargetLinker: A cytoscape app to integrate regulatory interactions in network analysis. PLoS One. 2013;**8**(12):e82160

[81] Bindea G, Galon J, Mlecnik B. CluePedia cytoscape plugin: Pathway insights using integrated experimental and in silico data. Bioinformatics. 2013;**29**(5):661-663

[82] van Iersel MP, Kelder T, Pico AR, Hanspers K, Coort S, Conklin BR, et al. Presenting and exploring biological pathways with PathVisio. BMC Bioinformatics. 2008;**9**:399

[83] van Iersel MP, Pico AR, Kelder T, Gao J, Ho I, Hanspers K, et al. The BridgeDb framework: Standardized access to gene, protein and metabolite identifier mapping services. BMC Bioinformatics. 2010;**11**:5

[84] Shen R, Olshen AB, Ladanyi M. Integrative clustering of multiple genomic data types using a joint latent variable model with application to breast and lung cancer subtype analysis. Bioinformatics. 2009;**25**(22): 2906-2912

[85] Rohart F, Gautier B, Singh A, Le Cao KA. mixOmics: An R package for 'omics feature selection and multiple data integration. PLoS Computational Biology. 2017;**13**(11):e1005752

[86] Meng C, Kuster B, Culhane AC, Gholami AM. A multivariate approach to the integration of multi-omics datasets. BMC Bioinformatics. 2014;**15**:162

[87] Wachter A, Beissbarth T. pwOmics: An R package for pathway-based integration of time-series omics data using public database knowledge. Bioinformatics. 2015;**31**(18):3072-3074

[88] McArdle S, Buscher K, Ehinger E, Pramod AB, Riley N, Ley K. PRESTO a new tool for integrating large-scale-omics data and discovering disease-specific signatures. bioRxiv. 2018:302604

[89] Chong J, Soufan O, Li C, Caraus I, Li S, Bourque G, et al. MetaboAnalyst 4.0: Towards more transparent and integrative metabolomics analysis. Nucleic Acids Research. 2018;**46**(W1):W486-W494

[90] Xia J, Fjell CD, Mayer ML, Pena OM, Wishart DS, Hancock RE. INMEX—A web-based tool for integrative meta-analysis of expression data. Nucleic Acids Research. 2013;**41**(Web Server issue):W63-W70

[91] Sharov AA, Schlessinger D, Ko MS. ExAtlas: An interactive online tool for meta-analysis of gene expression data. Journal of Bioinformatics and Computational Biology. 2015;**13**(6):1550019

[92] Hughey JJ, Butte AJ. Robust meta-analysis of gene expression using the elastic net. Nucleic Acids Research. 2015;**43**(12):e79

[93] Kim D, Li R, Dudek SM, Ritchie MD. ATHENA: Identifying interactions between different levels of genomic data associated with cancer clinical outcomes using grammatical evolution neural network. BioData Mining. 2013;**6**(1):23

[94] Carlin DE, Demchak B, Pratt D, Sage E, Ideker T. Network propagation in the cytoscape cyberinfrastructure. PLoS Computational Biology. 2017;**13**(10):e1005598

Section 3

Viral Systems Biology

Chapter 3

Lentiviral Vectors Come of Age? Hurdles and Challenges in Scaling Up Manufacture

Juan C. Ramirez

Abstract

The pharmaceutical industry has been attracted to the gene therapy field and is starting to support clinical trials, establishing collaborative strategies to develop commercial products which in many cases are based on lentiviral vectors. The predictable widespread use of lentiviral vectors in next-generation gene therapy scenarios aimed at dealing with not only rare diseases raises important challenges and hurdles regarding their manufacture. The author reflects on this in the chapter on the state of the art in the manufacture of lentiviral vectors, addressing some current manufacturing processes, their achievements, and the uncertainties in ensuring a validated process capable of releasing consistent vector quality that meets global health authorities' requirements. In summary, the proposal looks at the goals and challenges that must be addressed in manufacturing lentiviral vectors, in order to satisfy supply in the commercial stage, before we reach the next stage in gene therapy.

Keywords: lentiviral vector, large-scale manufacturing, gene therapy

1. Introduction

The practice of medicine is undergoing a revolution, moving from a focus on the treatment of symptoms, toward targeting the genetic cause of the disease. The huge development of disciplines, including but not restricted to molecular and cellular biology and genetic sciences, provides the framework for the advancement of individualized precision medicine. This new conception of medicine is based on the novel paradigm: the genes represent medicines themselves. Gene therapy is the groundbreaking strategy, which uses genes as medicines. Gene therapy is no longer an experimental approach [1, 2], and as with any novel therapy, patients' benefits must be balanced against the nonzero risk of the therapeutical approach. Most products currently assayed in clinical trials of gene therapy are viral vectors [3], i.e., biological products that challenge both the manufacturing processes in order to guarantee the supply of adequate quantities of the active vector and the regulatory requirements from the medicine agencies of target countries. In summary, viral vector production on a large scale implies novel challenges for a multidisciplinary field, in order to accommodate such specific requirements within the industrial process.

The first gene therapy experiments took advantage of the strategy that members of the *Retroviridae* family of viruses evolved to spread and remain stable in the host, with the integration of their genomes. Since those experiments we have assisted to a fast development of the viral vector field fuelled by promising data raised from early studies until the achievement of the current scenario [4, 5]. The gene therapy field is witnessing a sort of gold rush that is boosting personalized medicine by confronting many diseases as genetically treatable traits. Early vector developments focused on *Gammaretrovirus* as integrating entities for the delivery of a stable expression of the correctable gene, but since the last decade, they have been displaced by vectors deriving from the *Lentivirus* genus. As opposed to *Gammaretrovirus*, the *Lentivirus* displayed preferred integration sites in coding genomic regions rather than in transcription regulatory regions, and this has become a major safety feature to exploit. Nowadays, non-replicative self-inactivating lentiviral (SIN) vectors are used in the vast majority of novel gene therapy clinical trials using integrative vectors and are considered by now the optimal tools for ex vivo gene therapy and the safest and easiest-to-use vectors available for the delivery of genes into mammalian tissues [6].

The use of retroviral vectors in gene therapy as an emerging technology is following the Gartner hype cycle. Hope and expectation were seen when gene therapy entered the clinic in the early 1990s, but due to a lack of profound success and the unexpected death in 2000 of two patients, caused by the treatment, the expectations slowed down. In the following decade (2000–2010), two approaches coexisted: first, follow-up and ongoing clinical trials that were using an integrative type of vector (gammaretroviruses, gRV) used in those clinical trials resulting in unexpected fatal deaths and, second, intensive academic research focused on the development of new viral vectors and methodical exploration of the clinical procedure. This resulted in the advent of a novel type of vector called SIN-LV, derived from the causative agent of AIDS, the *Lentivirus* HIV, properly modified and engineered to render them safer (see Vectorizing HIV). In the current decade (2010 to the present), we have been recording investigative clinical trials using both vectors in several hematological and neurodegenerative rare diseases with a conclusion: the feasibility of the second-generation gene therapy approaches [4, 5]. This is mostly due to the huge development of SIN-LV vectors, due to their safer profile, in comparison with gammaretroviruses [6].

Gene therapy products have entered the commercialization phase, and a dozen treatments have been approved since 2012 in EU and the USA [1]. Up to last year, all of them were treatments for rare or ultrarare conditions, but in August 2017, a new key milestone for gene therapy development can be added to the chart, reinforcing the concept of gene therapy use in frequent pathologies in which current treatments are failing: the Novartis receives the first ever FDA approval for a CAR-T cell therapy, Kymriah® (tisagenlecleucel), for children and young adults with B-cell acute lymphoid leukemia (ALL). The goals reached in the past two decades in the gene therapy field open novel expectations offering a cure to rare genetic diseases, cancer, infectious diseases, and vaccine development, and in the short-medium term, innovations in the field will make affordable genetic intervention covering an array of diseases with a gene-defined cause [7].

The aim of this review is to recapitulate specific problems related to the manufacture of lentiviral vectors in particular during production stages, known as the upstream process (USP), focusing on the limitations that exist when scaling up, due to the nature of the virus, the particularities of the lentivector life cycle, and the producer cell line commonly used for production. There are excellent reviews [13–16] that provide a detailed description of the methodologies that can be followed in order to produce lentivectors on a large scale. There is a growing interest in the lentiviral vector field, and there are many topics worthy of a description. For those interested in more detailed, specific topics, i.e., manufacturing of CAR-T approaches, purification strategies, or specific problems linked to the target cell, I recommend references [10, 11, 17].

2. Delivering genes with *Lentivirus*

The virus-derived vector as delivery system lies at the heart of most currently employed forms of gene therapy; without the viral vectors, there is no treatment. These viruses must be custom-made in specialized facilities for each treatment, but manufacturing them is costly and onerous: it requires great expertise and multidisciplinary teams and specialized facilities with stringent conditions both for safety/containment and demanding production methods, under good manufacturing practice with regard to compliance (GMP) [8].

Viral vectors are complex bioproducts with an ordered architecture and are very sensitive to handling and environmental conditions. For these reasons, there are stringent requirements aimed at preserving biological activity during all stages of the manufacturing and delivery process. Then, during the production, purification, storage, and transportation stages, it is necessary to maintain specific rigorous control aimed at minimizing the loss of biological activity, in addition to controls which are common to other bioproducts, such as sterility. This implies that the manufacture of large amounts of a viral vector cannot simply be produced by transferring the know-how and well-stablished procedures developed in the pharmaceutical industry for the production of monoclonal antibodies or recombinant proteins.

The large-scale manufacture of lentivectors for use in humans is becoming the bottleneck in the success of ongoing or planned gene therapy development to be launched in the near future [9]. Indeed, the manufacturing capabilities of the companies to satisfy the short-medium term markets are central to decision-making for backers and investors, who are becoming cautious with regard to biotech firms developing gene therapy products that do not have a secure virus source [9]. Several papers have recently reported the need to succeed in developing a global manufacturing process for lentiviral vectors, driven by a deep understanding of both the product and the process, in order to establish the viral vector product profile and critical quality attributes [10, 11]. In addition, lentiviral products require the creation of a worldwide accepted and adopted international standard, suitable for the standardization of trials, in particular quantitation trials related to the potency of the target product, allowing a comparison of cross-manufacturing results for any lentiviral platform [12].

3. Vectorizing HIV

Human immunodeficiency virus (HIV) belongs to the family *Retroviridae*, subfamily *Orthoretrovirinae*, and genus *Lentivirus* of animal viruses. According to Baltimore's classification, it is an RNA reverse-transcribing virus (group VI). *Retroviridae* members are among the more exclusive entities in the virus taxon, and research in the field has provided outstanding insights into key concepts on biology, which were the basis demonstrating that dogmas do not stand up in science (the central dogma on molecular biology), providing one of the most useful tools in molecular biology (retrotranscriptase), supporting the concept of the existence of genes that provoke cancer (viral oncogenes), and finally, demonstrating that viral-genome integration also evolved in animal viruses as an efficient method of transmission. This breaking concept drove the original idea of gene therapy.

The HIV pandemic in the 1980s alerted humanity to the silent spreading of a deadly disease. HIV, a *Lentivirus*, was revealed to be a highly sophisticated virus with fine-tuned regulation, and it was mostly deciphered a decade after its discovery [5]. Scientists took advantage of the impressive knowledge gained about the virus in a decade transforming a dangerous virus into a safe viral vector. It is worth mentioning that HIV was described as a new virus in 1987, and barely 10 years later, a safe version of HIV-derived vectors was demonstrated as efficient in animal models [23].

Taking advantage of the previous studies with *Gammaretrovirus* vectors, the HIV-derived vectors were engineered to contain a mere 15–20% of the original virus, allowing plenty of room for transgenes/regulatory sequences. A method for production was established, and subsequent improvements ensure that the currently available format of the third generation of self-inactivating (SIN) vector [18] is generated as the safest and easiest-to-use vector available. It has multipurpose uses, from research and preclinical studies to clinical trials as there is a commercial product based on this type of vector. In all cases the procedure followed for production is almost the same, representing one of the main drawbacks, as manufacturing is not fully established, and a series of major concerns must be resolved in order to cover large mid- to long-term market requirements. Below is a brief summary of several relevant factors which must be addressed.

Production of HIV-derived vectors is a poorly optimized process, and a major hurdle to large-scale manufacturing is due in part to a deficient production of fully biologically active virions recovered in the culture media [19]. This is a dynamic process involving both production and inactivation rates, which ultimately renders on average 3–10 virus per producer cell [20], whereas during natural lentiviral infection, the number is close to 10^3/cell [21] and far removed from other vector systems such as AAV or *Adenovirus*, which render 10^4–10^5 viral particles per producer cell. There are at least four major issues during production that can result in this poor yield: (a) transfection robustness, (b) protein interactions during morphogenesis, (c) the nature of the cell system used for production, and (d) extremely labile essential components within the viral particles that lose activity during the production testing [22, 23].

3.1 DNA transfection

Production of HIV-derived lentivectors, and likewise other retroviral vectors derived from feline or equine *Lentivirus* or from gammaretroviruses, is based on DNA transfection of producer cells. The overall method was firstly demonstrated as feasible in pioneering research using poliovirus [24, 25], and it is based on the concept that the viral genome cloned in plasmids can recapitulate the genetic and morphogenetic instructions upon introduction in a eukaryotic cell in order to produce viral progeny. Early gene therapy studies developed a further step by splitting viral components in different plasmids, allowing the generation of non-replicative viral vectors as nonstructural/replication instructions which were no longer packaged in the progeny. These systems are currently also used to produce AAV-derived vectors. The basis is that packaging signals acting in cis are encoded in discrete regions of the viral genome. By including those sequences in the *transfer plasmid* bearing the therapeutic/reporter gene, transgenes are encapsidated in the virions. All the accessory functions are expressed during production from the so-called helper plasmids but are not licensed for encapsidation, as they do not carry the packaging signals, nor are they encoded in the transfer plasmid. The current model of production on large or small scales is based on DNA transfection of three or four plasmids. For a full, detailed description of the plasmid used in the production of

HIV-derived vectors, see [8]. As a result, efficiency is compromised by the proportion of cells transfected with the proper combination of plasmids, and indeed earlier second-generation production systems that use just three plasmids are more efficient in production [6].

3.2 Pseudotyping

Lentiviral vectors can be designed to carry heterologous envelope proteins. This pseudotyping allows the selective targeting of specific cells, conferring broader uses on the vector. Thus, lentivectors bearing RD114 [26], CD105 [27], and more recently measles virus [28] envelope glycoproteins, among others, have been described as conferring specific targeting in B cells, T cells, and hematopoietic stem cells, respectively. However, most of the lentivectors that have entered into the clinic and are commonly being used in research are pseudotyped with *Vesicular stomatitis* virus g protein (VSVg). There are two reasons for this. First, a wide number of different cell types are targeted by such an envelope, and second, it confers robustness on the viral particle during the purification stages [29].

However, the presence of the VSV receptor in the producer cell line contributes to diminishing the viral burden in the harvest [30]. Envelope proteins are membrane proteins that pass through the secretory pathway involving the endoplasmic reticulum cisternae and the Golgi apparatus, before they reach the plasma membrane, a system also used for the synthesis and recycling of the membrane receptors. Prevention of a premature encounter between ligand and receptor is mandatory, in order to increase the env protein available for the morphogenetic program. Indeed, the impact of this phenomenon has evolved in the natural infection of HIV. Vpu, an HIV accessory protein (see below), plays a dual role in the viral cycle, firstly by promoting egress in a cell-type-dependent manner and secondly by controlling the recircularization of envelope proteins during the synthesis and preventing premature binding of the HIV gp160 env protein with the CD4 natural receptor during T-cell infection [31, 32]. To our knowledge no data has been published which accounts for the impact of such a process (autotransduction) during HIV-derived VSV-pseudotyped processes, but data obtained in our company indicate that this phenomenon is actually taking place in the producer cell. By specific quantitation of retrotranscribed RNA and integrated copies of cDNA in the producer cell, we have been able to quantitate that 30–50% loss of the actual viral particles produced are lost by reentering in the producer cell. Different lines are currently being developed at VIVEbiotech to minimize or fully prevent this phenomenon.

3.3 Helper functions

In the case of HIV, unlike to *Gammaretrovirus*, there is an array of six viral products collectively named accessory proteins. Their roles are different throughout the viral life cycle. Some are related to controlling the innate/cellular immune response (vif), others modulate the adaptive immune response (nef), while others are mandatory for a fully regulated genetic program (tat, rev), morphogenesis and the egress (vpu), or viral infectivity (vif, nef).

HIV-derived vectors are produced in the absence of five of the six accessory proteins encoded by the wild-type virus: with the exception of rev, the other five are not expressed in helper plasmids. The rationale is to minimize the presence of viral sequences in the transfer plasmid, enabling safer vectors and minimizing the recombination between viral sequences in order to limit the risk of rescue of wild-type virus during production. Tat protein was unnecessary within the design of third-generation vectors, as no viral promoter was used in production. However, as has been demonstrated, the effect of those other accessory proteins is not negligible, and their function during vector production is controversial [31, 32].

3.4 Manufacturing virus from DNA

As described above DNA transfection is the current and unique manner to produce lentiviral vectors. Three major concerns must be considered about this approach:

- *Efficiency*. Transfection of four plasmids raises a few concerns that affect the reproducibility and efficacy [6]. Alternative procedures rely on the generation of stable producer cells bearing helper functions, limiting the transfection to just the transfer plasmid, which aids robustness and production yields. Different approaches for either constitutive or inducible systems have been designed [33], but in most cases low titers have been achieved. Toxicity of the VSV proteins has been cited as limiting more efficient systems.

- *Quantity/quality*. DNA, just like any other reagent employed in the manufacturing process, requires identical strict compliance with GMP. The production of batches requires a certification of analysis that includes protein contamination, sterility, and the sequencing of all plasmids, both helper and transfer plasmids. These tests and the large amount of DNA required to scale-up the process require an improvement in cost-effective methods regarding quantity.

- *Scalability*. As mentioned above, several transfection procedures are currently available, but only two are relevant to large-scale manufacturing, whether using calcium phosphate and/or PEI (polyethylenimine) as matrixes to accomplish stable DNA complexes (currently proposed as methods for scale-up manufacturing). It is worth pointing out that to our knowledge there are no systematic studies addressing the relative efficiency of the two methods, taking into account not just vector yields but also cost-effectiveness and the impact on the cost of the production process.

4. Manufacturing *Lentivirus*: the VIVEbiotech approach

There are several excellent reviews on the specific steps during the manufacture of lentiviral vectors, the approaches to consider when scaling up and the critical points to consider for decision-making [13, 14, 34]. However, all of the processes must conciliate at least the next three considerations: (a) potency, meaning the capability of producing large quantities of vector; (b) robustness, i.e., highly reproducible; and (c) versatility, as demand changes according to project needs.

Manufacturing follows a process of production, purification, and concentration. At every step, specific features of the lentiviral vectors must be taken into account. Critical concerns to be considered include:

1. *Production*. It is worth remembering that lentiviral vectors are enveloped, and both cellular proteins and lipid content can vary depending on the culture conditions [22]. As previously mentioned, pseudotyping is of major relevance, as it can affect the fate of the produced virions and also contribute to the

physical properties of the particles [35] and interaction with the solid phase in bioreactors.

2. *Harvesting*. It requires a consideration of the fact of interaction of the virions with the solid phase in bioreactors, in addition to the highly unstable nature of lentivectors, with their very short half-life [19], and a high dependence of biological activity on physical conditions and particle content [36–38].

3. *Purification*. It is critical for maintenance of the bona fide conditions of the genome and capsid, and unfortunately lentivectors are currently purified by chemically based procedures, such as ion-exchange chromatography, which significantly affect particle viability. To the author's knowledge, there is no current industrial process for purification using affinity chromatography.

Herein is an overview of the manufacturing process developed at VIVEbiotech, which focuses on some of the critical steps. VIVEbiotech has obtained authorization from the European Medicines Agency (EMA) to provide cGMP batches of lentiviral vectors under the manufacturing process the company has fully developed (see **Figure 1**). VIVEbiotech is currently releasing batches of lentiviral vectors produced under these conditions to clients in EU and the USA. VIVEbiotech's manufacture of lentivectors is based on the fixed-bed bioreactor iCELLis™ supplied by Pall. High compaction in the solid phase allows for a large culture surface, ranging from 0.53 to 2.6 m^2 (Nano™ configuration) in 1 liter disposable bioreactor, and is further scalable up to approximately 300 m^2 (500 + ™ configuration) based on chromatography principles. In the 1 liter small configuration, the carriers made of PET are fixed in a 40 mL chamber, and the process is monitored by probes controlling cell growth, and physical–chemical conditions are monitored by in-process BioXpert software. The harvest is collected by perfusion and purified by ion-exchange chromatography. It is then concentrated by tangential flow filtration, rendering a final concentration factor close to 300-fold and reaching a yield on a per-surface basis in accordance with market standards [39]. This process has been fully optimized in two remarkable steps:

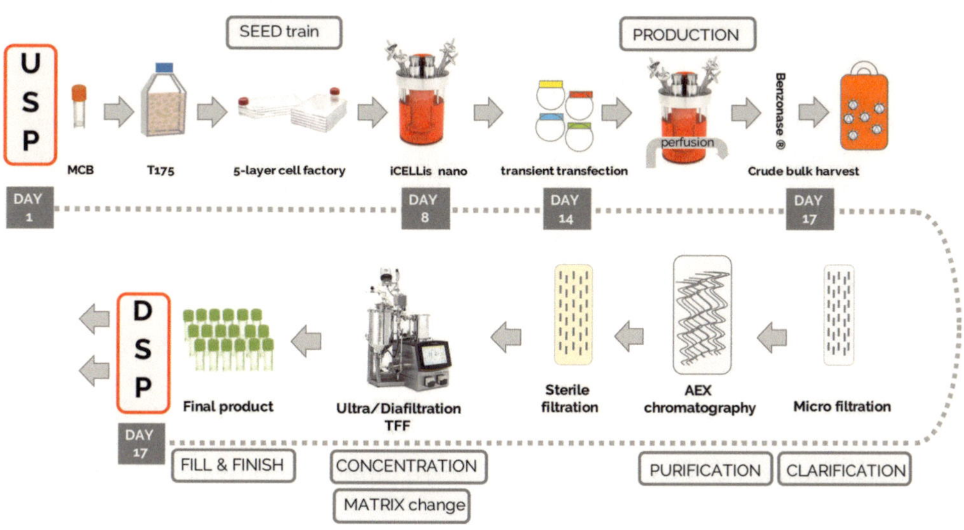

Figure 1.
Manufacturing process of lentiviral vectors optimized at VIVEbotech.

- The seed train of producer HEK293T cells has been optimized in order to minimize the cell density of the seed. The conditions are the lowest standard achieved for other cell types which (i) allow growth with almost no operator intervention; minimizing the risk contamination, (ii) growth is fully monitored by the BioXpert program, ensuring constant, reliable growth conditions, and (iii) this has allowed us to shorten the process by a third.

- DNA transfection is based on DNA-calcium-phosphate precipitation, making our process highly reliable (close to 90% efficiency) and cost-effective, even under cGMP.

5. Challenging by numbers

As mentioned above, vector production is becoming a roadblock which is hitting gene therapy capabilities. Let us examine a few numbers in order to understand the size of the problem. So, what does lentivector-based gene therapy need to do in order to ensure it can be applied in the future? Lentiviral vectors for what? In their current design, lentivectors are capable of being used in the treatment of blood disorders, central nervous system disorders, immune therapy for certain cancers, and neurological conditions that can be treated with stem cells delivering a cargo of corrector genes. How can this be transformed into numbers? Certain statements require understanding, before setting out the main points which need to be addressed.

Leaving aside the fact that for every condition treatable by gene therapy, the number of patients is highly variable; estimations can be made using a highly prevalent disease under phase III by bluebird bio (www.bluebird.com) such as beta-thalassemia/sickle-cell trait (SCT). According to NIH data, 1100 infants are born every year among the African American community, and more than 100,000 individuals are estimated to have SCD in the USA; in Africa 15 million Africans are estimated to have SCD, and there are 200–300,000 affected births per year worldwide (https://www.ncbi.nlm.nih.gov/books/NBK1377/). Current conditions for transduction efficiently into hematopoietic stem cells require around $>1 \times 10^{10}$ biologically active particles per vector dose per infant patient though these data can vary depending on specific features of each treatment. According to current standards of the manufacturing process to achieve such production, harvest produced from 2 square meters and equivalent to 4–6 liters of harvest per patient would need to be produced, representing a huge quantity of 10,000 liters per year to treat new infant cases in the USA for SCD. Current platforms of production and, significantly, purification procedures are not capable of addressing this situation.

6. Conclusions

Gene therapy is no longer an experimental approach to treat genetic diseases. Several medicine agencies worldwide have approved the commercialization of medicinal products based on viral vectors as intermediate medicinal products. This raises the need to manufacture large quantities of viral vectors under costly cGMP manufacturing environment. There are a limited number of pharmaceutical and biotechnology companies capable to manufacture and release lentiviral vectors of defined composition and quality control in quantities to attend the foreseeable market needs. Challenges for the development of more controlled and cost-effective manufacturing process have yet to be overcome. The complicated manufacturing process needs to be simplified to promote standardization and yield products of increased defined composition. However, there are still open questions that arise from the system employed for production, principally related to the model of production based on DNA-transfected produced cells. Automatization of the manufacturing process is also required, in order to increase capabilities leading to an industrialization process. This will contribute to developing global manufacturing processes for lentiviral vectors and help to establish the target product profile and quality attributes. In summary, efforts in modifying the current manufacturing model of lentivectors are needed to facilitate the entry into commercialization stages.

Acknowledgements

The dedication, hard work, and helping discussions of all the team working at VIVEbiotech are strongly acknowledged. The author wants to give special thanks to the cGMP producer team Maite Azcarate and Janire Arizeta for their excellent work and continuous effort to develop the manufacturing process of the company. The meticulous work of the QA/QP management by Estibaliz Perez and Rakel Lopez de Maturana is also acknowledged.

The projects developed at VIVEbiotech have been partially defrayed by project RTC-2015-3393-1 by the Spanish Ministry of Economy.

Conflict of interest

The author declares no conflict of interest.

Author details

Juan C. Ramirez
VIVEbiotech, Donostia/San Sebastian, Spain

*Address all correspondence to: jcramirez@vivebiotech.com

References

[1] BioPharma Dealmakers. Gene therapy ready to deliver. Nature Reviews Drug Discovery. 2017;**16**:B24-B25

[2] Kumar SRP, Markusic DM, Biswas M, High KA, Herzog RW. Clinical development of gene therapy: Results and lessons from recent successes. Molecular Therapy. Methods & Clinical Development. 2016;**3**:16034:1-11. DOI: 10.1038/mtm.2016.34

[3] Kotterman MA, Chalberg TW, Schaffer DV. Viral vectors for gene therapy: Translational and clinical outlook. Viral Vectors for Gene Therapy. 2015;**17**:63-89. DOI: 10.1146/annurev-bioeng-071813-104938

[4] Dunbar CE, High KA, Joung JK, Kohn DB, Ozawa K, Sadelain M. Gene therapy comes of age. Science. 2018;**359**:1-10. DOI: 10.1126/science.aan4672

[5] Naldini L, Trono D, Verma IM. Lentiviral vectors, two decades later. Science. 2016;**353**:1101-1102. DOI: 10.1126/science.aah6192

[6] White M, Whittaker R, Gandara C, Stoll EA. A guide to approaching regulatory considerations for lentiviral-mediated gene therapies. Human Gene Therapy Methods. 2017;**28**:163-176. DOI: 10.1089/hgtb.2017.096

[7] Mavilio F. Developing gene and cell therapies for rare diseases: An opportunity for synergy between academia and industry. Gene Therapy. 2017;**24**:590-592. DOI: 10.1038/gt.2017.36

[8] Gandara C, Affleck V, Stoll EA. Manufacture of third-generation lentivirus for preclinical use, with process development considerations for translation to good manufacturing practice. Human Gene Therapy Methods. 2018;**29**:1-15. DOI: 10.1089/hgtb.2017.098

[9] Kolata G. Gene therapy's strange roadblock. The New York Times. 2017:D12

[10] Levine BL, Miskin J, Wonnacott K, Keir C. Global manufacturing of CAR T cell therapy. Molecular Therapy. Methods & Clinical Development. 2017;**4**:92-101. DOI: 10.1016/j.omtm.2016.12.006

[11] Wang X, Isabelle Rivière J. Clinical manufacturing of CAR T cells: Foundation of a promising therapy. Molecular Therapy—Oncolytics. 2016;**3**:16015. DOI: 10.1038/mto.2016.15

[12] Zhao Y, Stepto H, Schneider CK. Development of the first World Health Organization lentiviral vector standard: Toward the production control and standardization of lentivirus-based gene therapy products. Human Gene Therapy Methods. 2017:205-214. DOI: 10.1089/hgtb.2017.078

[13] Merten OW, Scheweizer M, Chahal P, Karmen AA. Manufacturing of viral vectors: Part I. Upstream processing. Pharmaceutical Bioprocessing. 2014;**2**:183-202. DOI: 10.4155/PBP.1416

[14] Merten OW, Scheweizer M, Chahal P, Karmen AA. Manufacturing of viral vectors: Part II. Downstream processing and safety aspects. Pharmaceutical Bioprocessing. 2014;**2**:237-251. DOI: 10.4155/PBP.1415

[15] Schweizernd M, Merten OW. Large-scale production means for the manufacturing of lentiviral vectors. Current Gene Therapy. 2010;**10**:474-486

[16] Merten OW, Charrier S, Laroudie N, Fauchille S, Dugue S, Jenny C, et al. Large-scale manufacture and characterization of a lentiviral vector produced for clinical ex vivo gene therapy application. Human Gene Therapy. 2011;**22**:343-356. DOI: 10.1089/hum.2010.060

[17] Bandeira V, Peixoto C, Rodrigues AF, Cruz PE, Alves PM, Coroadinha AS, et al. Downstream processing of lentiviral vectors: Releasing bottlenecks. Human Gene Therapy Methods. 2011;**23**:255-263. DOI: 10.1089/hgtb.2012.059

[18] Zufferey R, Dull T, Mandel RJ, et al. Self-inactivating lentivirus vector for safe and efficient in vivo gene delivery. Journal of Virology. 1998;**72**: 9873-9880

[19] Thomas JA, Ott DE, RJ G. Efficiency of HIV post-entry infection process: Evidence against disproportionate numbers of defective virus. Journal of Virology. 2007:4367-4370

[20] Andreadis S, Lavery T, Davis H, Le Doux JM, Yarmush ML, Morgan JR. Toward a more accurate quantitation of the activity of recombinant retroviruses: Alternatives to titer and multiplicity of infection. Journal of Virology. 2000;**74**:1258-1266

[21] Chen YH, Di Mascio M, Perelson AS, Ho DD, Zhang L. Determination of virus burst size in vivo using a single-cycle SIV in rhesus macaques. PNAS. 2007;**104**:19079-19084. DOI: 10.107/pnas.0707449104

[22] Carmo M, Alves A, Rodrigues AF, Coroadinha AS, Carrondo MJT, Alves PM, et al. Stabilization of gammaretroviral and lentiviral vectors: From production to gene transfer. The Journal of Gene Medicine. 2009;**11**:670-678. DOI: 10.1002/jgm.1353

[23] Coroadinha AS, Ribeiro J, Rolda A, Cruz PE, Alves PM, Merten OW, et al. Effect of medium sugar source on the production of retroviral vectors for gene therapy. Biotechnology and Bioengineering. 2006;**94**:24-36. DOI: 10.1002/bit.20778

[24] http://www.sciencemag.org/news/2002/07/poliovirus-baked-scratch

[25] Cello J, Paul AV. Wimmer E chemical synthesis of poliovirus cDNA: Generation of infectious virus in the absence of natural template. Science. 2002;**297**:1016-1018. DOI: 10.1126/science.1072266

[26] Bell AJ Jr, Fegen D, Ward M, Bank A. RD114 envelope proteins provide an effective and versatile approach to pseudotype lentiviral vectors. Experimental Biology and Medicine. 2010;**235**:1269-1276. DOI: 10.1258/ebm.2010.010053

[27] Vanden Driessche T, Chuah MK. Targeting endothelial cells by gene therapy. Blood. 2013;**122**:1993-1994. DOI: 10.1182/blood-2013-08-518266

[28] Levy C, Amirache F, Girard-Gagnepain A, Frecha C, Roman-Rodrıguez FJ, Bernadin O, et al. Measles virus envelope pseudotyped lentiviral vectors transduce quiescent human HSCs at an efficiency without precedent. Blood Advances. 2017;**1**:2088-2104. DOI: 10.1182/bloodadvances.2017007773

[29] Frecha C, Lévy C, Cosset F-L, Verhoeyen E. Advances in the field of lentivector-based transduction of T and B lymphocytes for gene therapy. Molecular Therapy. 2010;**18**:1748-1757. DOI: 10.1038/mt.2010.178

[30] Ohishi M, Shioda T, Sakuragi J-i. Retro-transduction by virus pseudotyped with glycoprotein of vesicular stomatitis virus. Virology. 2007;**362**:131-138. DOI: 10.1016/j.virol.2006.12.030

[31] Damme NV, Guatelli J. HIV-1 Vpu inhibits accumulation of the envelope glycoprotein within clathrin-coated, Gag-containing endosomes. Cellular Microbiology. 2008;**10**:1040-1057. DOI: 10.1111/j.1462-5822.2007.01101

[32] Neil SJD, Eastman SW, Jouvenet N, Bieniasz PD. HIV-1 Vpu promotes release and prevents endocytosis of

nascent retrovirus particles from the plasma membrane. PLoS Pathogens. 2006;**2**:354-367. DOI: 10.1371/journal.ppat.0020039.g011

[33] Manceur AP, Kim H, Misic V, Andreev N, Dorion-Thibaudeau J, Lanthier S, et al. Scalable lentiviral vector production using stable HEK293SF producer cell lines. Human Gene Therapy Methods. 2017;**28**:330-339. DOI: 10.1089/hgtb.2017.086

[34] Ausubel LJ, Hall C, Sharma A, Shakeley R, Lopez P, Quezada V, et al. Production of cGMP-grade lentiviral vectors. Bioprocess International. 2012;**10**(2):32-43

[35] Richieri SP, Bartholomew R, Aloia RC, Savary J, Gore R, Halt J, et al. Characterization of highly purified, inactivated HIV-1 particles isolated by anion exchange chromatography. Vaccine. 1998;**16**:119-129

[36] Le Doux JM, Davis HE, Morgan JR, Yarmush ML. Kinetics of retrovirus production and decay. Biotechnology and Bioengineering. 1999;**63**:654-662

[37] Carmo M, Panet A, Carrondo MJT, Alves PM, Cruz PE. From retroviral vector production to gene transfer: Spontaneous inactivation is caused by loss of reverse transcription capacity. The Journal of Gene Medicine. 2008;**10**:383-391. DOI: 10.1002/jgm.1163

[38] Carmo M, Dias JD, Panet A, Coroadinha AS, Carrondo MJT, Alves PM, et al. Thermosensitivity of the reverse transcription process as an inactivation mechanism of lentiviral vectors. Human Gene Therapy. 2009;**20**:1168-1176. DOI: 10.1089=hum.2009.068

[39] Valkama AJ, Leinonen HM, Lipponen EM, Turkki V, Malinen J, Heikura T, et al. Optimization of lentiviral vector production for scale-up in fixed-bed bioreactor. Gene Therapy. 2018;**25**:39-46. DOI: 10.1038/gt.2017.91

Chapter 4

Orf Virus: A New Class of Immunotherapy Drugs

Ruixue Wang and Shuhong Luo

Abstract

Orf is one of the most widespread viral diseases worldwide, usually benign and self-limiting, and mainly affects not only sheep and goats, but also various other ruminants and mammals. The causative agent, orf virus (ORFV) is a member of the genus parapoxvirus, owing to its zoonotic importance and ability to cross-infect other species sporadically. ORFV encodes virulence and immunomodulatory factors that interfere with host inflammatory effect and antiviral immune mechanisms and induces a transient and complex cytokine response, initially represented by Th1-related cytokines followed by Th2-related cytokines. The ORFV has evolved several mechanisms to survive in the presence of the immune system, resulting in repeated infections. Currently, ORFV has been developed as vaccines in veterinary field. The unique host immune escape ability obtained by ORFV has made it one of the important candidates for prevention and treatment of various diseases (including chronic viral diseases, tumor, and liver fibrosis).

Keywords: immunoregulation, immunotherapy, orf virus, vaccine, oncolysis virus

1. Introduction

Orf was first discovered in Europe in 1920. At present, the main geographical distribution of the pathogen is not clear, which is considered to be prevalent worldwide since orf exists in all areas where sheep exists. The United States of America, Germany, Korea, Japan, India, Argentina, Malaysia, Egypt, and China have reported the occurrence and prevalence of the disease, which has brought a certain degree of loss to the sheep industry. The causative agent, orf virus (ORFV), also known as contagious pustular stomatitis (contagious ecthyma) virus, belongs to the parapoxvirus genus of poxvirus family and causes nonsystemic cutaneous disease by mainly infecting sheep and goats. In recent years, the cases of human, camel, yak, red squirrel, cat, domestic reindeer, etc. infected by ORFV have been reported. This indicates that the host range of the virus is expanding.

After being infected by ORFV, the infected animals begin with the appearance of erythema on the lips, tongue, nose, and breast of sheep, then develop into papules, blisters, and pustules, and finally form crusts, characterized by proliferative inflammation (**Figure 1**). The course of the disease is mostly an acute infection, healed within 1-2 months, but there are also cases of chronic persistent infection records. The disease rarely causes animal death unless host immunosuppression or secondary infection occurs, but there are also reports of a high mortality rate of 93% in young goats.

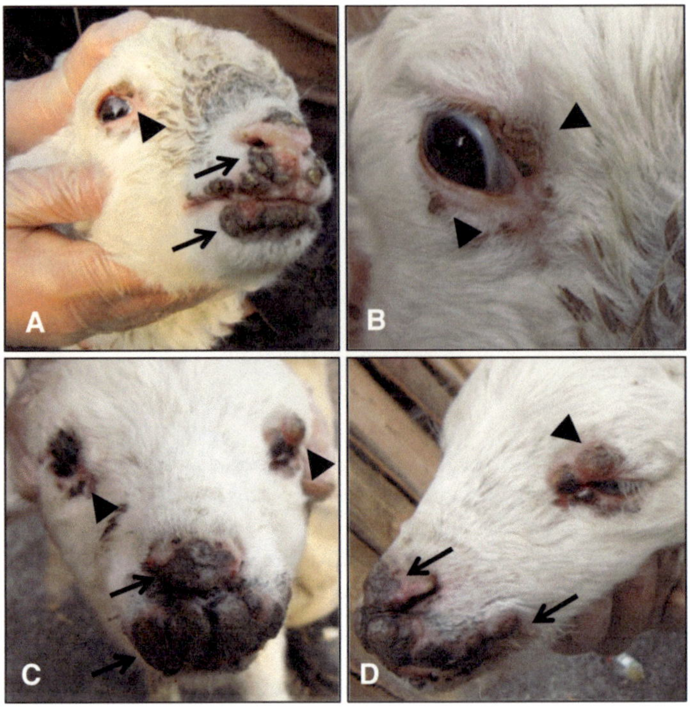

Figure 1.
Typical clinical signs of ORFV infection in sheep (cited from reference [1]). Proliferative skin lesions of two infected sheep around the mouth, nostrils (arrows), and the eyelids (arrowheads).

2. ORFV genome biology

ORFV belongs to the subfamily parapoxvirus of the poxvirus family. Other members of this genus include pseudovaccinia virus (PCPV), bovine papular stomatitis virus (BPSV), and parapoxvirus of red deer in New Zealand (PVNZ). The mature ORFV particles are 250–280 mm in length and 170–200 nm in width, and elliptical and coiled shaped, while the immature virus particles are conical, brick-shaped, and special coiled spherical particles (**Figure 2**). The surface of the virus particles showed a characteristic braided helical structure of cross-arranged around the long axis of the virus particles for eight-shaped winding. There are other ways of winding, and the virus particles encapsulated outside the capsule. ORFV replicates and matures in the cytoplasm, encoding polymerases associated with virus replication and transcription.

ORFV is a linear double-stranded (ds) DNA virus with a genome size of 134–139 kb. The average G + C content of the virus genome was approximately 64%. The content of G + C in the ORFs in the central coding region of the genome is not very different, but the ORFs in the two ends of the genome are very different, which are even less than 50% in some regions, such as ORFV127. Mercer believes that the terminal is based on the conservation of the genome sequence (OVSA00, OVIA82, and NZ2) and the transcriptional initiation of the gene in three strains of the virus. Even though the G + C content of seven ORFVs in ORFV102–104, 109–112 is quite different from the genome average G + C content, and the homology of the encoded proteins is low, but the G + C content between different strains is very close [2]. The content of G + C in BPSV genome terminal variant region is similar, too, which is a marker of poxvirus members. Like other members of the poxvirus family, it has a large central coding region in the middle of the genome and an inverted terminal repeat (ITR) at both ends, which had covalently closed terminal hairpin structures (**Figure 3**).

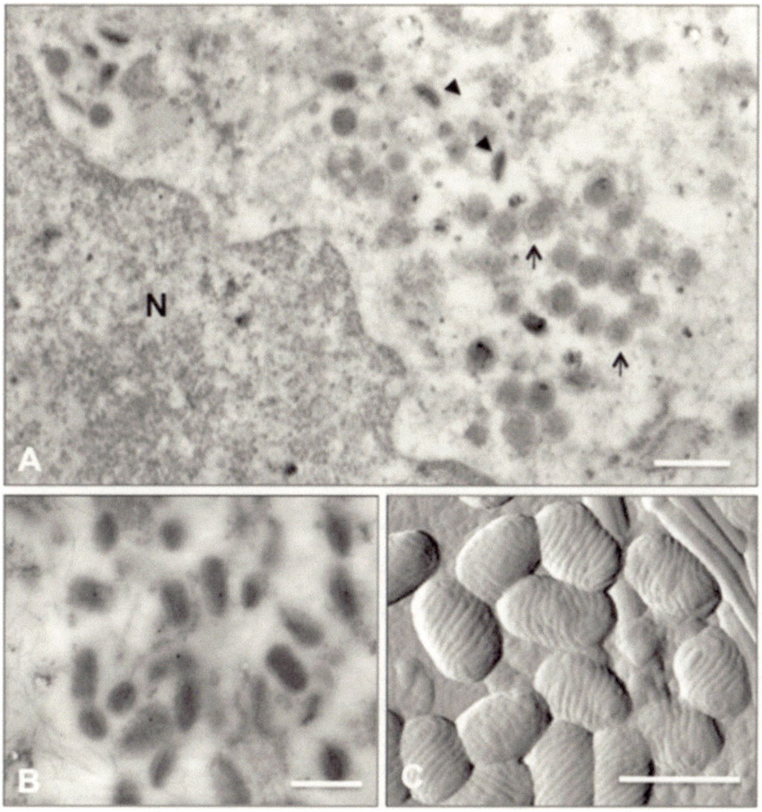

Figure 2.
The electron microscopy of the ORFV (cited from reference [1]). The predominantly immature virions (arrows and arrowheads) (A) and intracellular mature virions (B) under transmission electron microscopy. (C) The extracellular virions under atomic force microscopy. Scale bars = 500 nm.

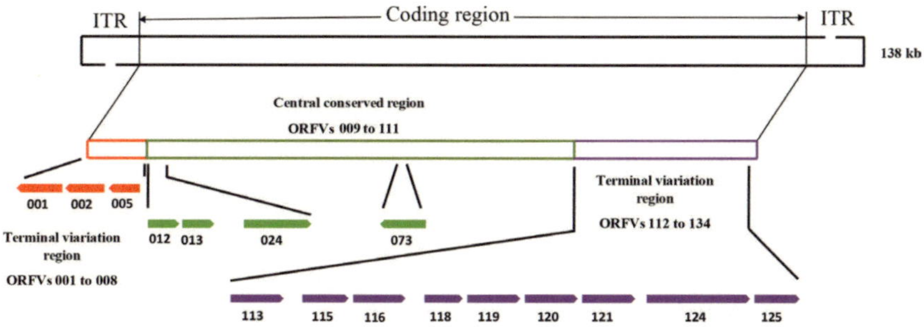

Figure 3.
The structure of the genome of ORFV strain NA1/11. The genome of ORFV is 138 kb, encoding 132 genes, and includes highly variable terminal regions, responsible for virus virulence and pathogenesis, and relatively conserved central regions with a high GC content for viral replication and virus morphogenesis. There are 16 novel genes unique to parapoxvirus, with putative virulence-host range functions.

There are many studies on the whole gene sequencing and analysis. The first genome of two strains (SA00 and IA82) was sequenced in 2003 by the Mei Dao Animal Disease Center of the US Department of Agriculture. OVSA00 was identified as the reference sequence of the virus. Mercer submitted the full genome sequence of the NZ2 strain in 2006, and McGuire sequenced strain (D1701) and submitted it to GenBank. Luo submitted the full genome sequence of the ORFV

Strain	Species of origin	Country of origin	No. of predicted genes	Genome size (bp)	ITR size (bp)	G + C content	Geninfo identifier	Reference
D1701	Sheep	Germany	288	134,038			325073632	[3]
YX	Goat	China	132	138,231	3446	63.8	913203877	[4]
NP	Goat	China	124	132,111	2426	63.8	913204143	[4]
GO	Goat	China	132	139,886	3964	63.6	913204010	[4]
SJ1	Goat	China	129	139,112	4153	63.6	913204268	[4]
HN3/12	Sheep	China	132	136,643	2794	63.7	1240685690	[2]
SA00	Goat	USA	132	139,962	3936	63.4	40019123	[5]
NA1/11	Sheep	China	134	137,080	3020	63.6	632123481	[6]
IA82	Sheep	USA	132	137,241	3092	64.3	40019122	[5]
NZ2	Sheep	New Zealand	132	137,820	3389	64.3	74230714	[7]

Table 1.
Summary of complete genomic sequence data of 10 ORFV strains.

strains of China. Up to now, NCBI has included 10 complete genome sequences of ORFV (**Table 1**).

Delhon et al. conservatively estimated about 130 coding genes in the ORFV genome [5]. Through analysis of the genome sequence of NZ2 strain, 132 possible coding genes were found in the genome of this virus strain. Transcription initiation elements (TAAAT) existed before the coding regions of the two genes and were found in BPSV. Similar conserved sequences also exist, but these two genes only exist in parapoxvirus but not in orthopoxvirus. Mercer et al. checked the ORFs of 24 genes of ORFV, which showed high interspecific variability mainly in the two terminal variant regions [7]. Many genes located in the core region have been identified. The ORFV050 gene, similar to L4R of vaccinia virus (VACV), encodes the DNA-binding virion core protein VP8 [8]. ORFV057 encodes protein OH1, analogous to the VACV structural protein VH1, that can dephosphorylate phosphatidylinositol 3, 5-bisphosphate, and plays a role in virion maturation [9]. The ORFV011 (B2L) gene, a homolog of the F13 L gene of VACV, encodes a major envelope protein of 42 kDa, which is thought to be a lipase. Additionally, the viral A32L gene (ORFV108) encodes an ATPase involved in virion DNA packaging [10]. Virulence genes, coding genes related to host pathogenesis and immunoregulatory genes are located in the ITR regions of the ORFV genome, such as ORFVs 007, 020, 112, 117, 119, 125, and 132.

3. The immunomodulatory ability of ORFV

3.1 ORFV and immunomodulatory ability

ORFV is widely recognized as having a powerful host immunoregulatory function. ORFV can quickly mediate humoral and adaptive immune responses. After being infected with ORFV, many cells of the innate immune system are activated and induce the secretion of chemokines and cytokines. Neutrophils, natural killer (NK) cells, and dendritic cells (DC) are recruited at the site of infection. In the early stage of infection, ORFV mainly induces the Th1-type immune response. Peripheral immune cells secrete IFN-γ, TNF-a, IL-6, IL-8, IL-12, IL-18, and then Th2-type immune response appears, mainly inducing secreting of IL-4, IL-10, IL-1 receptor antagonists (IL-1RA). The conditioning of complement and antigen-presenting-cell (APC)-mediated antigen presentation are important steps to activate the immune response.

ORFV and some of its encoded proteins have a good immunomodulatory function. Homologous alignment analysis of the host sequence and the viral gene sequence has identified that some ORFV genes have corresponding immunoregulatory functions, including coding for IL-10 homologous proteins, chemokine binding proteins, secretory inhibitors of GM-CSF and IL-2, vascular endothelial factor (VEGF), and interferon resistance protein. The main targets of IFN resistance genes are host cytokines, chemokines, NF-κB signaling pathway, and apoptosis pathway. The synergistic effect of these proteins has strong immunomodulatory effects on ORFV.

ORFV is the only virus that contains the gene encoding lL-10 found in the poxvirus family. vIL-10, a 21.7 kDa protein with remarkable homology to IL-10, which is encoded by ORFV127 gene, plays an important role in immunosuppression through inhibition of cytokine syntheses, such as TNF-α and IL-8, IFN-γ [11], suppression of the maturation and functionality of DC [12, 13], blockage of Th1 cell activation indirectly through weakening the antigen processing, and presentation ability of APC. The direct role of the vIL-10 gene in virulence was demonstrated using an ORFV lacking the IL-10 gene, which showed attenuated properties in animal experiments [14], while vIL-10 can exert immunostimulatory effects by inducing moderate compensatory immune activation [15].

One of the characteristics of ORFV infection is the proliferation of capillaries and the increasing of permeability in the dermis, which is caused by viral VEGF. The deletion of VEGF gene leads to vascular permeability reduction, inhibition of epidermal cells and inactivation of VEGF receptors (VEGF-2) [16, 17]. Viral VEGF, sharing 16–27% of amino acid identity with its homolog VEGFs (VEGF-A, VEGF-B, VEGF-C, VEGF-D), has the same function as VEGFs: promoting the proliferation of epidermal cells, inducing the proliferation of host vascular endothelial cells and increasing the permeability of capillary vessel wall. Interestingly, VEGF variants are observed in different strains of ORFV. Despite such variation existing, the functional domains of the protein exhibit conserved structure. Studies reported that a recombinant virus strains lacking VEGF gene reduced vascular changes characteristic of natural infections, with less proliferation of blood vessels and dermal edema, pustule, and scab formation in ORFV pathogenesis [18].

In the early stage of ORFV infection, the orf virus interferon resistance protein (OVIFNR) encoded by ORFV20 binds to the viral replication intermediates and prevents the termination of IFN-induced virus-carrying protein translation. OVIFNR shares 31% sequence similarity with the E3L protein of VACV, and the C-terminal region with the binding activity of dsRNA (or viral replication intermediates) was the necessary region to prevent the antiviral activity of IFN and associated with pathogenicity and host tropism. OVIFNR eliminates the antiviral effect of IFN through the synthesis of the dsRNA-dependent protein, like protein kinase (PKR) [19]. The dsRNA-activated PKR is one of the main antiviral proteins induced by IFN. Activated PKR phosphorylates the translation initiation factor eIF2-α and impairs protein synthesis to inhibit viral replication [20]. OVIFNR not only compete with PKR to bind to viral replication intermediates but also inhibit the activity of PKR, thus preventing host cell interference from terminating the translation of viral proteins.

The ORFV007-encoded dUTPase clusters with mammalian counterparts and is more similar to mammalian dUTPases than to dUTPases from other poxviruses [21]. The virulence of ORFV with the 007-gene deletion is significantly lower than that of natural ORFV.

Chemokine-binding protein (CBP), the coding product of ORFV112, has similarities in structure and function with CBP II of orthopoxvirus and rabbit poxvirus. It can bind and inhibit chemokine and prevent chemokine-receptor interaction. When chemokines bind to their receptor, the G protein-coupled receptor, the white blood cells were recruited and activated in viral infection. In addition, chemokines interact with glycosaminoglycan (GAG) and establish a gradient liquid phase that guides leukocytes through the endothelium into tissues. ORFV-CBP and CBP-II have high affinity to some CC chemokines, such as monocyte chemoattractant protein 1 (MCP1), macrophage inflammatory protein 1 (MIP1), and regulated on activation, normal T cell expressed and secreted (RANTES), which can produce chemotactic effects of both nucleus/macrophage and T lymphocyte toward inflammation. Although ORFV encodes a number of secreted anti-inflammatory factors, the deletion of the CBP gene severely attenuated viral virulence and pathogenesis [22].

GIF is encoded by ORFV117, expressed in the late stage of infection, and has the dual activity of inhibiting host GM-CSF and IL-2, thus inhibiting host immune activity [23]. The gene is conserved in different ORFV strains and also exists in other parapoxvirus strains, but the amino acid sequence similarity between ORFV and BPSN is only 40%. However, the function of GIF in virulence and pathogenesis is not yet known.

3.2 ORFV and immune evasion

After infection, sheep produces antibodies to four or five immunodominant antigens [24–26]. Murine monoclonal antibodies recognizing 42 kDa envelope proteins, the 10 kDa putative fusion protein, and 65 kDa antigens have been described that can discriminate between the different parapoxvirus species [27, 28]. In spite of an apparently normal immune response to infection, sheep can be repeatedly infected, suggesting that, in common with other large DNA viruses, ORFV has evolved an immune evasion strategy [29, 30].

ORFV infection stimulated hyperplasia of epidermal cells and capillaries growth with increased vascular permeability, which allows increased virus replication and formation of scabs on wound healing. Scabs are rich in virus particles and provide temporary refuge for viruses to escape from immunization. The antiviral effect of IFN is the first line of defense against viral infection; ORFV evades immune clearance by inhibiting IFN-stimulated genes expression mediated by the JAK/STAT signaling pathway [31]. In addition, ORFV also can induce apoptosis mediated by CD95 pathway [15] or inhibit the pro-inflammatory NF-κB signaling, a crucial regulator of host innate immune responses. For pathogens, interfering with the activation of NF-κB is a particular strategy against host defense mechanisms. The regulation of NF-κB includes the regulation of IκB in the cytoplasm, and post-translational phosphorylation, acetylation, and methylation in the nucleus. The ORFV 002, 024, 073, 119, and 121 genes have been reported that play roles in NF-κB pathway regulation [32–38].

ORFV002 is an early and late stage virus gene, mainly located in the nucleus. ORFV002-encoded protein can inhibit the activation of NF-κB pathway induced by TNF-α and ORFV virus infection, which may through interfering the interaction between NF-κB-p65 and P300 in the nucleus block the acetylation of NF-κB-p65 Lys310 when phosphorylation occurs at ORFV002 Ser 276 [32, 33]. The 52 amino acids of ORFV002 N terminal may interact with protein S100A4 [34].

The ORFV024-encoded protein combines with LAGE3 to inhibit the phosphorylation of IKKs complex and then affects the phosphorylation of NF-κB-p65, inhibits the host immune cells to secrete some important cytokines, and regulates the host's immune response [35].

ORFV119 blocks the NF-κB signaling largely in a pRb-dependent manner, by inhibiting IKK complex activation early in infection [36]. ORFV119 interacted with TNF receptor-associated factor 2 (TRAF2), an adaptor protein recruited to signaling complexes upstream of IKK in infected cells, in a LxCxE motif-dependent manner, which leads to inhibition of NF-κB signaling.

The ORFV073 protein, 188 amino acids with a molecular weight of 21.8 kDa, whose protein at 149, 160, and 166 locations contained three predicted and partially overlapping nuclear localization signals, is located in nucleus during viral replication and be related to gene expression regulation. When the ORFV073 gene was deleted in the ORFV genome, the expression of chemokines and other pro-inflammatory genes was significantly increased, and most of the gene expression changes were regulated by the NF-κB transcription factor family [37].

After infected with an ORFV121 gene deletion mutant, NF-κB-mediated gene transcription was increased, while the expression of ORFV121 in cell cultures significantly decreased NF-κB-regulated reporter gene expression, suggesting that NF-κB inhibitor binds to and inhibits the phosphorylation and nuclear translocation of NF-κB-p65 in the cell cytoplasm, thus providing a mechanism for the inhibition of NF-κB-p65 phosphorylation and nuclear translocation [38].

4. Diagnosis and treatment

At present, there is no international standard for the diagnosis of amniotic aphthous ulcer, mainly based on the typical clinical symptoms and laboratory tests to diagnose. As shown in **Figure 4**, current laboratory diagnoses include PCR, ELISA, electron microscopy, histopathology, Western blotting.

The orf is mainly observed by the morphology of virus particles. A human case of orf was identified by transmission electron microscopy in a 20-year-old woman with two painful pruritic lesions on her left index finger [39]. Under the transmission electron microscopy, multiple typical ORFV particles existed with brick-shaped morphology, consisting of a central DNA-containing core surrounded by a bilayered capsid.

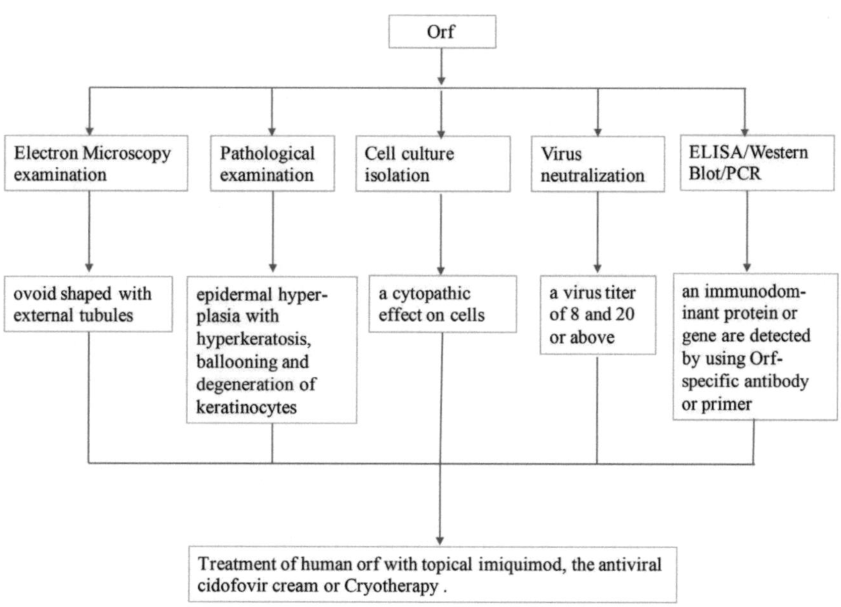

Figure 4.
The diagnosis methods and treatment of orf.

Histopathological features of the orf lesion include vacuole and swelling of keratinocytes, interstitial degeneration, marked epidermal hyperplasia, micro-swelling in the epidermis, aggregation of neutrophils, DC, T cells and B cells in the subcutaneous tissue, and formation of the crust. Eosinophilic inclusion bodies were also evident in the infected cytoplasm, but not at all stages of infection. There are mononuclear cells infiltrating into the dermis, such as phagocytes, lymphocytes, and eosinophils. In secondary infection, a dermal infiltration of neutrophils appeared [40].

A variety of primary cells and cell lines can be used to isolate ORFV. Primary lamb testicular cells and primary lamb kidney-trophoblast cells, which were originally used by Plowright et al., are the most common in the prozonal cells [41]. The primary fetal lamb muscle cells [42] and turbinate cells [5], fetal bovine lung cells [43], Madin-Darby bovine kidney cells [44], and vero cells [45] were also used. ORFV can induce cytopathy in these cells after inoculation of the first generation or blindly transmitted for one to two generations. The common cytopathic effects are aggregation, fusion, shrinkage, and cell detachment. Eosinophils in the cytoplasm can be observed by staining of infected cell lines. Parapoxvirus culturing, in general, is considered to be difficult, with a need for many passages before observing cytopathic effects, such as ballooning, wounding, degeneration of cells, etc.

Serum neutralization tests, commonly used for the detection of antibodies, are not considered to be the method of choice for primary diagnosis, as immunity to ORFV is mainly cell-mediated, and neutralizing antibodies are usually at small concentrations. Suspected serum with a titer of 8 and 20 or above is considered as positive for orf infection, in serum neutralization test and complement-fixation test, respectively [46].

Indirect ELISA with a purified virus as coating antigen, peroxidase complex protein A, G and chimeric A/G as secondary antibodies can be used to detect antibody levels in different animals [47]. The method has been successfully applied to detect the virus in camels [48], lambs [49], and humans [50] suspected to be infected with the disease.

A 40 kDa immunogenic protein has been found in the positive sera of infected animals by Western blotting. Similarly, two proteins of approximately 22 and 20 kDa have been found by this method.

Different PCR methods can be used for rapid diagnosis of ORFV. A conventional PCR assay based on amplification of the ORFV B2L gene, a homolog of the F13 L gene of VACV, encodes a major envelope protein of 42 kDa, which is supposed to be a lipase, and was used to detect parapoxvirus species. A duplex PCR assay using A29 gene (413 bp) and H3L gene (708 bp) has the potential to differentiate capripoxviruses from ORFV [51]. A single-step PCR method was applied for the rapid differential diagnosis of ORFV infections [52]. Primers targeting the A32L gene, besides, the complete sequences of another two viral genes were also investigated: the B2L, and E3L genes, which encodes a dsRNA-binding protein. A conventional PCR assay combined with DNA sequencing can be used to distinguish among the different parapoxvirus species [53]. A sensitive and specific SYBR Green I real-time PCR assay was performed to quantitatively detect ORFV [54].

Restriction fragment length polymorphism (RFLP) of the genome is a powerful tool for analyzing the molecular characteristics of poxvirus, which can potentially distinguish different strains of amniotic stomatitis virus. Restriction enzyme fragments are obtained by enzyme (EcoRI, BamHI, and HindIII) digestion. Commercialized kits with random amplified polymorphic DNA have been used to distinguish virus strains from large numbers of species. Loop-mediated isothermal amplification targeting B2L, DNA polymerase, and F1L genes have been developed and proven to be effective diagnostic tools [55, 56].

The ORFV has a strong ability to adapt to the external environment and has a strong resistance, even after a year in the sheep pen around, the virus still has a strong infectivity. But the virus is sensitive to temperature changes, suitable for the humid environment, and can be killed at 60°C in 30 min. Besides, the use of antiviral drugs has been applied in human and animal orf infections with satisfying results, such as Cidofovir [57, 58].

The virus is virulent for about a year when it is added to 50% glycerol saline and stored at 4°C. In the actual feeding process, generally choosing to use 20% hot grass and wood ash solution, 10% lime milk, 2% sodium hydroxide solution, and 1% acetic acid for disinfection can kill the virus, while using 2% sodium hydroxide solution to kill the virus in 5 min. The treatment for human orf is often focused on secondary infection. Previous reports have described speeding up the healing process with topical imiquimod [59] and the antiviral cidofovir cream [60]. Cryotherapy has also been used successfully to treat orf cases, especially in immunocompromised patients [61].

5. Clinical applications

ORFV is widely recognized as a virus with powerful host immunoregulatory function, but neutralizing antibody in ORFV infection is rare [27], and passive transfer of antibody-rich colostrum or serum does not protect lambs from infection [62, 63].

In veterinary medicine, ORFV is widely used as a new type of preventive and immunomodulatory preparation. Activated or inactivated ORFV preparations have a dose-dependent immunomodulatory function. Therefore, based on ORFV, drugs for a variety of infectious animal diseases have been developed (Baypamun; Zylexis). For example, Baypamun was used to suppress stress-related infections in horses, and clinical data showed a significant 40% reduction in the incidence of stress-related infections in the medication group [64]. Its therapeutic effect has also been verified in other animals, such as the treatment of bovine herpesvirus type 1, chronic stomatitis or infectious peritonitis in cats, and breast tumors in dogs. Inactivated ORFV can induce spontaneous regulation of cytokine responses in mice, such as up-regulation of Th1 cytokines (IL-12, IL-18, and IFN-γ), activation of CD14 and TLR-mediated monocyte activation, and release of anti-inflammatory Th2-related cytokines.

5.1 Antiviral preparations

Activated or inactivated ORFV makes many kinds of animals to fight different viral diseases. The ORFV has significant antifibrous activity in CCL4-mediated liver fibrosis [65]. The inactivated ORFV agents with a low dose (only 500,000 virus particles) are more effective in transgenic mice than the standard 3TC for HBV infection [66]; thus, it can be used as the candidate antiviral agent for the treatment of human HBV. Inactivated ORFV has anti-HCV activity in vitro and transgenic mice model [67]. In addition, ORFV can prevent the recurrence of fatal herpes simplex virus (HSV) and recurrence of genital herpes in Guinea [68].

ORFV can be used as a carrier to produce new animal recombinant vaccines. Recombinant ORFV vector induces an antiviral response in various animals. Recombinant pseudorabies virus glycoprotein gC or gD can be used to prevent infection in mice and pigs [69, 70]. ORFV recombinant with protein P40 can induce the immune protection of rats from infection, and effectively eliminated the Borna disease virus in the brain [71]. ORFV recombinant classical swine fever virus protein E2 can also make pigs immune to classical swine fever virus (CSFV) [72]. Recombinant ORFV expressing hemagglutinin (HA) or nucleoprotein (NP) of highly pathogenic avian influenza virus H5N1 protects mice against H5N1 and H1N1 influenza viruses [73]. A recombinant virus strain D1701-VP1 of rabbit blast virus VP60 gene expression induced the infected cells releasing goblet-like particles to protect the rabbit from ORFV attack [74].

5.2 Oncolytic virus

Live or inactivated ORFV induces antitumor immune responses in multiple tumor models. Fiebig et al. reported for the first time that inactivated ORFV has antitumor effects in a variety of tumor metastasis models, such as mouse-transplanted malignant melanoma B16F10 and human breast cancer MDA-MB-231 models [75], and found that NK cells play an important role in the antitumor of ORFV. After neutralizing IFN-γ, the antitumor effect disappeared, while the anti-NK-1.1 antibody partially weakened the antitumor activity of ORFV by inhibiting the activity of NK and NKT cells. Inactivated ORFV inhibited tumor growth in a mouse MDA-MB-231 tumor model without NK and lacking functional T and B lymphocytes. Whether inactivated or active ORFV is used to treat mouse tumor models, NK cells play an important role in antitumor. A study by Rintoul further confirmed that ORFV inhibits tumor growth of melanoma and colorectal cancer, and proved that ORFV could play an antitumor role by activating NK cells and stimulating their secretion of cytokines IFN-γ and granzyme B [76]. Tai et al. found that surgery mediated the dysfunction of NK cells [77]. Intraoperative injection of ORFV improves the function of NK cells, thereby reducing intraoperative metastasis and prolong survival. Recently, a study of the virus strain CF189, which is the high similarity with the ORFV virus strain NZ2 obtained by homologous recombination, showed that CF189 effectively kills three negative breast cancer cells with time and dose dependence [78].

6. Conclusion

ORFV causes orf, a nonsystemic, highly contagious, ubiquitous disease of sheep and goats [79], which is characterized by maculopapular and proliferative lesions affecting the skin around the mouths, nostrils, and teats. Virus virulence and immunomodulation genes of ORFV contribute to combat local inflammatory response, innate immunity (including apoptosis, NK cell activation, and antiviral response), and immune adaptation. Therefore, ORFV has been used in veterinary medicine as preventive and therapeutic immunomodulatory agents. Moreover, live or inactivated ORFV preparations exhibit immunomodulatory effects, with therapeutic efficacy demonstrated for various diseases, including infectious diseases and tumors.

Acknowledgements

This work was partially supported by grants from the National Natural Science Foundation of China (NSFC) (nos. 81773271 and 31672536), the Foshan University Senior Talent Start Fund (20161110004, y2016-ky69, and Gg07043), the

Foshan University High-level University Fund (y2016-td148, 20170131020), and Guangdong Provincial Education Department of Education of Guangdong Province (2017KZDXM088). The funders had no role in study design, data collection and analysis, decision to publish, or preparation of the manuscript.

Conflict of interest

The authors declare that no conflict of interest exists.

Author details

Ruixue Wang[1,2] and Shuhong Luo[1*]

1 Department of Laboratory Medicine, School of Stomatology and Medicine, Foshan University, Foshan, Guangdong Province, P.R. China

2 Department of Basic Medical Sciences, School of Stomatology and Medicine, Foshan University, Foshan, Guangdong Province, P.R. China

*Address all correspondence to: sluo815@gmail.com

References

[1] Li W, Ning Z, Hao W, Song D, Gao F, Zhao K, et al. Isolation and phylogenetic analysis of orf virus from the sheep herd outbreak in northeast China. BMC Veterinary Research. 2012;**8**:229. DOI: 10.1186/1746-6148-8-229

[2] Chen H, Li W, Kuang Z, Chen D, Liao X, Li M, et al. The whole genomic analysis of orf virus strain HN3/12 isolated from Henan province, central China. BMC Veterinary Research. 2017;**13**(1):260. DOI: 10.1186/s12917-017-1178-1

[3] McGuire MJ, Johnston SA, Sykes KF. Novel immune-modulator identified by a rapid, functional screen of the parapoxvirus ovis (Orf virus) genome. Proteome Science. 2012;**10**(1):4. DOI: 10.1186/1477-5956-10-4

[4] Chi X, Zeng X, Li W, Hao W, Li M, Huang X, et al. Genome analysis of orf virus isolates from goats in the Fujian province of southern China. Frontiers in Microbiology. 2015;**6**:1135. DOI: 10.3389/fmicb.2015.01135

[5] Delhon G, Tulman ER, Afonso CL, Lu Z, de la Concha-Bermejillo A, Lehmkuhl HD, et al. Genomes of the parapoxviruses ORF virus and bovine papular stomatitis virus. Journal of Virology. 2004;**78**(1):168-177. DOI: 10.1128/JVI.78.1.168-177.2004

[6] Li W, Hao W, Peng Y, Duan C, Tong C, Song D, et al. Comparative genomic sequence analysis of Chinese orf virus strain NA1/11 with other parapoxviruses. Archives of Virology. 2015;**160**(1):253-266. DOI: 10.1007/s00705-014-2274-1

[7] Mercer AA, Ueda N, Friederichs SM, Hofmann K, Fraser KM, Bateman T, et al. Comparative analysis of genome sequences of three isolates of Orf virus reveals unexpected sequence variation. Virus Research. 2006;**116**(1-2):146-158. DOI: 10.1016/j.virusres.2005.09.011

[8] Wang H, Jiang J, Ding R, Wang X, Liao M, Shao J, et al. Identification and characterization of Orf virus 050 protein proteolysis. Virus Genes. 2017;**53**(3):400-409. DOI: 10.1007/s11262-017-1430-6

[9] Segovia D, Haouz A, Porley D, Olivero N, Martinez M, Mariadassou M, et al. OH1 from Orf virus: A new tyrosine phosphatase that displays distinct structural features and triple substrate specificity. Journal of Molecular Biology. 2017;**429**(18):2816-2824. DOI: 10.1016/j.jmb.2017.07.017

[10] Yogisharadhya R, Bhanuprakash V, Venkatesan G, Balamurugan V, Pandey AB, Shivachandra SB. Comparative sequence analysis of poxvirus A32 gene encoded ATPase protein and carboxyl terminal heterogeneity of Indian orf viruses. Veterinary Microbiology. 2012;**156**(1-2):72-80. DOI: 10.1016/j.vetmic.2011.10.021

[11] Imlach W, McCaughan CA, Mercer AA, Haig D, Fleming SB. Orf virus-encoded interleukin-10 stimulates the proliferation of murine mast cells and inhibits cytokine synthesis in murine peritoneal macrophages. The Journal of General Virology. 2002;**83**(Pt 5):1049-1058. DOI: 10.1099/0022-1317-83-5-1049

[12] Chan A, Baird M, Mercer AA, Fleming SB. Maturation and function of human dendritic cells are inhibited by orf virus-encoded interleukin-10. The Journal of General Virology. 2006;**87**(Pt 11):3177-3181. DOI: 10.1099/vir.0.82238-0

[13] Lateef Z, Fleming S, Halliday G, Faulkner L, Mercer A, Baird M. Orf virus-encoded interleukin-10 inhibits maturation, antigen presentation

and migration of murine dendritic cells. The Journal of General Virology. 2003;**84**(Pt 5):1101-1109. DOI: 10.1099/vir.0.18978-0

[14] Fleming SB, Anderson IE, Thomson J, Deane DL, McInnes CJ, McCaughan CA, et al. Infection with recombinant orf viruses demonstrates that the viral interleukin-10 is a virulence factor. The Journal of General Virology. 2007;**88**(Pt 7):1922-1927. DOI: 10.1099/vir.0.82833-0

[15] Kruse N, Weber O. Selective induction of apoptosis in antigen-presenting cells in mice by Parapoxvirus ovis. Journal of Virology. 2001;**75**(10):4699-4704. DOI: 10.1128/JVI.75.10.4699-4704.2001

[16] Savory LJ, Stacker SA, Fleming SB, Niven BE, Mercer AA. Viral vascular endothelial growth factor plays a critical role in orf virus infection. Journal of Virology. 2000;**74**(22):10699-10706. DOI: 10.1128/JVI.74.22.10699-10706.2000

[17] Ueda N, Wise LM, Stacker SA, Fleming SB, Mercer AA. Pseudocowpox virus encodes a homolog of vascular endothelial growth factor. Virology. 2003;**305**(2):298-309. DOI: 10.1006/viro.2002.1750

[18] Wise LM, Savory LJ, Dryden NH, Whelan EM, Fleming SB, Mercer AA. Major amino acid sequence variants of viral vascular endothelial growth factor are functionally equivalent during Orf virus infection of sheep skin. Virus Research. 2007;**128**(1-2):115-125. DOI: 10.1016/j.virusres.2007.04.018

[19] Tossici-Bolt L, Fleming JS, Conway JH, Martonen TB. An analytical technique to recover the third dimension in planar imaging of inhaled aerosols--2 estimation of the deposition per airway generation. Journal of Aerosol Medicine. 2007;**20**(2):127-140. DOI: 10.1089/jam.2007.0577

[20] Li H, Zhu X, Zheng Y, Wang S, Liu Z, Dou Y, et al. Phylogenetic analysis of two Chinese orf virus isolates based on sequences of B2L and VIR genes. Archives of Virology. 2013;**158**(7):1477-1485

[21] Cottone R, Buttner M, McInnes CJ, Wood AR, Rziha HJ. Orf virus encodes a functional dUTPase gene. The Journal of General Virology. 2002;**83**(Pt 5):1043-1048. DOI: 10.1007/s00705-013-1641-7

[22] Fleming SB, McCaughan C, Lateef Z, Dunn A, Wise LM, Real NC, et al. Deletion of the chemokine binding protein gene from the parapoxvirus Orf virus reduces virulence and pathogenesis in sheep. Frontiers in Microbiology. 2017;**8**:46. DOI: 10.3389/fmicb.2017.00046

[23] Deane D, McInnes CJ, Percival A, Wood A, Thomson J, Lear A, et al. Orf virus encodes a novel secreted protein inhibitor of granulocyte-macrophage colony-stimulating factor and interleukin-2. Journal of Virology. 2000;**74**(3):1313-1320. DOI: 10.1128/JVI.74.3.1313-1320.2000

[24] McKeever DJ, Jenkinson DM, Hutchison G, Reid HW. Studies of the pathogenesis of orf virus infection in sheep. Journal of Comparative Pathology. 1988;**99**(3):317-328. DOI: 10.1016/0021-9975(88)90052-7

[25] Yirrell DL, Reid HW, Norval M, Howie SE. Immune response of lambs to experimental infection with Orf virus. Veterinary Immunology and Immunopathology. 1989;**22**(4):321-332. DOI: 10.1016/0165-2427(89)90168-2

[26] Chand P, Kitching RP, Black DN. Western blot analysis of virus-specific antibody responses for capripox and contagious pustular dermatitis viral infections in sheep. Epidemiology and Infection. 1994;**113**(2):377-385. DOI: 10.1017/S0950268800051803

[27] Czerny CP, Waldmann R, Scheubeck T. Identification of three distinct antigenic sites in parapoxviruses. Archives of Virology. 1997;**142**(4):807-821. DOI: 10.1007/s007050050120

[28] Housawi FM, Roberts GM, Gilray JA, Pow I, Reid HW, Nettleton PF, et al. The reactivity of monoclonal antibodies against orf virus with other parapoxviruses and the identification of a 39 kDa immunodominant protein. Archives of Virology. 1998;**143**(12):2289-2303. DOI: 10.1007/s007050050461

[29] Alcami A, Koszinowski UH. Viral mechanisms of immune evasion. Immunology Today. 2000;**21**(9):447-455. DOI: 10.1016/S0167-5699(00)01699-6

[30] Haig DM. Subversion and piracy: DNA viruses and immune evasion. Research in Veterinary Science. 2001;**70**(3):205-219. DOI: 10.1053/rvsc.2001.0462

[31] Harvey R, McCaughan C, Wise LM, Mercer AA, Fleming SB. Orf virus inhibits interferon stimulated gene expression and modulates the JAK/STAT signalling pathway. Virus Research. 2015;**208**:180-178. DOI: 10.1016/j.virusres.2015.06.014

[32] Ning Z, Zheng Z, Hao W, Duan C, Li W, Wang Y, et al. The N terminus of orf virus-encoded protein 002 inhibits acetylation of NF-kappaB p65 by preventing Ser(276) phosphorylation. PLoS One. 2013;**8**(3):e58854. DOI: 10.1371/journal.pone.0058854

[33] Diel DG, Luo S, Delhon G, Peng Y, Flores EF, Rock DL. A nuclear inhibitor of NF-kappaB encoded by a poxvirus. Journal of Virology. 2011;**85**(1):264-275. DOI: 10.1128/JVI.01149-10

[34] Chen D, Zheng Z, Xiao B, Li W, Long M, Chen H, et al. Corrigendum: Orf virus 002 protein targets ovine protein S100A4 and inhibits NF-kappaB signaling. Frontiers in Microbiology. 2017;**8**:160. DOI: 10.3389/fmicb.2017.00160

[35] Diel DG, Delhon G, Luo S, Flores EF, Rock DL. A novel inhibitor of the NF-{kappa}B signaling pathway encoded by the parapoxvirus orf virus. Journal of Virology. 2010;**84**(8):3962-3973. DOI: 10.1128/JVI.02291-09

[36] Nagendraprabhu P, Khatiwada S, Chaulagain S, Delhon G, Rock DL. A parapoxviral virion protein targets the retinoblastoma protein to inhibit NF-kappaB signaling. PLoS Pathogens. 2017;**13**(12):e1006779. DOI: 10.1371/journal.ppat.1006779

[37] Khatiwada S, Delhon G, Nagendraprabhu P, Chaulagain S, Luo S, Diel DG, et al. A parapoxviral virion protein inhibits NF-kappaB signaling early in infection. PLoS Pathogens. 2017;**13**(8):e1006561

[38] Diel DG, Luo S, Delhon G, Peng Y, Flores EF, Rock DL. Orf virus ORFV121 encodes a novel inhibitor of NF-kappaB that contributes to virus virulence. Journal of Virology. 2011;**85**(5):2037-2049. DOI: 10.1371/journal.ppat.1006561

[39] Peng F, Chen Z, Zheng SY, Li HM, Du J, Zhang JZ. A case of Orf identified by transmission electron microscopy. Chinese Medical Journal. 2016;**129**(1):108-109. DOI: 10.4103/0366-6999.172606

[40] Vikoren T, Lillehaug A, Akerstedt J, Bretten T, Haugum M, Tryland M. A severe outbreak of contagious ecthyma (orf) in a free-ranging musk ox (Ovibos moschatus) population in Norway. Veterinary Microbiology. 2008;**127**(1-2):10-20. DOI: 10.1016/j.vetmic.2007.07.029

[41] Plowright W, Ferris RD. Studies with rinderpest virus in tissue culture.

II. Pathogenicity for cattle of culture-passaged virus. Journal of Comparative Pathology. 1959;**69**(2):173-184

[42] McInnes CJ, Wood AR, Nettleton PE, Gilray JA. Genomic comparison of an avirulent strain of Orf virus with that of a virulent wild type isolate reveals that the Orf virus G2L gene is non-essential for replication. Virus Genes. 2001;**22**(2):141-150

[43] Inoshima Y, Murakami K, Wu D, Sentsui H. Characterization of parapoxviruses circulating among wild Japanese serows (Capricornis crispus). Microbiology and Immunology. 2002;**46**(8):583-587. DOI: 10.1111/j.1348-0421.2002.tb02738.x

[44] Guo J, Zhang Z, Edwards JF, Ermel RW, Taylor CJ, de la Concha-Bermejillo A. Characterization of a North American orf virus isolated from a goat with persistent, proliferative dermatitis. Virus Research. 2003;**93**(2):169-179. DOI: 10.1016/S0168-1702(03)00095-9

[45] Tedla M, Berhan N, Molla W, Temesgen W, Alemu S. Molecular identification and investigations of contagious ecthyma (Orf virus) in small ruminants, North west Ethiopia. BMC Veterinary Research. 2018;**14**(1):13. DOI: 10.1186/s12917-018-1339-x

[46] Zarnke RL, Dieterich RA, Neiland KA, Ranglack G. Serologic and experimental investigations of contagious ecthyma in Alaska. Journal of Wildlife Diseases. 1983;**19**(3):170-174. DOI: 10.7589/0090-3558-19.3.170

[47] Inoshima Y, Shimizu S, Minamoto N, Hirai K, Sentsui H. Use of protein AG in an enzyme-linked immunosorbent assay for screening for antibodies against parapoxvirus in wild animals in Japan. Clinical and Diagnostic Laboratory Immunology. 1999;**6**(3):388-391

[48] Azwai SM, Carter SD, Woldehiwet Z. Immune responses of the camel (*Camelus dromedarius*) to contagious ecthyma (Orf) virus infection. Veterinary Microbiology. 1995;**47**(1-2):119-131. DOI: 10.1016/0378-1135(95)00055-F

[49] McKeever DJ, Reid HW, Inglis NF, Herring AJ. A qualitative and quantitative assessment of the humoral antibody response of the sheep to orf virus infection. Veterinary Microbiology. 1987;**15**(3):229-241

[50] Yirrell DL, Vestey JP, Norval M. Immune responses of patients to orf virus infection. The British Journal of Dermatology. 1994;**130**(4):438-443. DOI: 10.1016/0378-1135(87)90077-0

[51] Zheng M, Liu Q, Jin N, Guo J, Huang X, Li H, et al. A duplex PCR assay for simultaneous detection and differentiation of Capripoxvirus and Orf virus. Molecular and Cellular Probes. 2007;**21**(4):276-281. DOI: 10.1016/j.mcp.2007.01.005

[52] Chan KW, Hsu WL, Wang CY, Yang CH, Lin FY, Chulakasian S, et al. Differential diagnosis of orf viruses by a single-step PCR. Journal of Virological Methods. 2009;**160**(1-2):85-89. DOI: 10.1016/j.jviromet.2009.04.025

[53] Inoshima Y, Morooka A, Sentsui H. Detection and diagnosis of parapoxvirus by the polymerase chain reaction. Journal of Virological Methods. 2000;**84**(2):201-208. DOI: 10.1016/S0166-0934(99)00144-5

[54] Wang Y, Yang K, Bai C, Yin D, Li G, Qi K, et al. Development of a SYBR Green I real-time PCR for the detection of the orf virus. AMB Express. 2017;7(1):21. DOI: 10.1186/s13568-016-0322-9

[55] Tsai SM, Chan KW, Hsu WL, Chang TJ, Wong ML, Wang CY. Development of a loop-mediated isothermal

amplification for rapid detection of orf virus. Journal of Virological Methods. 2009;**157**(2):200-204. DOI: 10.1016/j.jviromet.2009.01.003

[56] Li J, Song D, He W, Bao Y, Lu R, Su G, et al. Rapid detection of orf virus by loop-mediated isothermal amplification based on the DNA polymerase gene. Archives of Virology. 2013;**158**(4):793-798. DOI: 10.1007/s00705-012-1526-1

[57] De Clercq E. Cidofovir in the therapy and short-term prophylaxis of poxvirus infections. Trends in Pharmacological Sciences. 2002;**23**(10):456-458. DOI: 10.1016/S0165-6147(02)02091-6

[58] Dal Pozzo F, Andrei G, Lebeau I, Beadle JR, Hostetler KY, De Clercq E, et al. In vitro evaluation of the anti-orf virus activity of alkoxyalkyl esters of CDV, cCDV and (S)-HPMPA. Antiviral Research. 2007;**75**(1):52-57. DOI: 10.1016/j.antiviral.2006.11.010

[59] Erbagci Z, Erbagci I, Almila Tuncel A. Rapid improvement of human orf (ecthyma contagiosum) with topicalimiquimod cream: Report of four complicated cases. The Journal of Dermatological Treatment. 2005;**16**:353-356

[60] De Clercq E. Clinical potential of the acyclic nucleoside phosphonates cidofovir, adefovir, and tenofovir in treatment of DNA virus and retrovirus infections. Clinical Microbiology Reviews. 2003;**16**:569-596. DOI: 10.1128/CMR.16.4.569-596.2003

[61] Degraeve C, De Coninck A, Senneseael J, Roseeuw D. Recurrent contagious ecthyma (Orf) in an immunocompromised host successfully treated with cryotherapy. Dermatology. 1999;**198**:162-163. DOI: 10.1159/000018095

[62] Buddle BM, Pulford HD. Effect of passively-acquired antibodies and vaccination on the immune response to contagious ecthyma virus. Veterinary Microbiology. 1984;**9**(6):515-522. DOI: 10.1016/0378-1135(84)90013-0

[63] Mercer AA, Yirrell DL, Reid HW, Robinson AJ. Lack of cross-protection between vaccinia virus and orf virus in hysterectomy-procured, barrier-maintained lambs. Veterinary Microbiology. 1994;**41**(4):373-382. DOI: 10.1016/0378-1135(94)90033-7

[64] Horohov DW, Breathnach CC, Sturgill TL, Rashid C, Stiltner JL, Strong D, et al. In vitro and in vivo modulation of the equine immune response by parapoxvirus ovis. Equine Veterinary Journal. 2008;**40**(5):468-472. DOI: 10.2746/042516408X322111

[65] Nowatzky J, Knorr A, Hirth-Dietrich C, Siegling A, Volk HD, Limmer A, et al. Inactivated Orf virus (Parapoxvirus ovis) elicits antifibrotic activity in models of liver fibrosis. Hepatology Research. 2013;**43**(5):535-546. DOI: 10.1111/j.1872-034X.2012.01086.x

[66] Weber O, Siegling A, Friebe A, Limmer A, Schlapp T, Knolle P, et al. Inactivated parapoxvirus ovis (Orf virus) has antiviral activity against hepatitis B virus and herpes simplex virus. The Journal of General Virology. 2003;**84**(Pt 7):1843-1852. DOI: 10.1099/vir.0.19138-0

[67] Paulsen D, Urban A, Knorr A, Hirth-Dietrich C, Siegling A, Volk HD, et al. Inactivated ORF virus shows antifibrotic activity and inhibits human hepatitis B virus (HBV) and hepatitis C virus (HCV) replication in preclinical models. PLoS One. 2013;**8**(9):e74605. DOI: 10.1371/journal.pone.0074605

[68] Ons E, Van Brussel L, Lane S, King V, Cullinane A, Kenna R, et al. Efficacy of a Parapoxvirus ovis-based immunomodulator against equine herpesvirus type 1 and

Streptococcus equi equi infections in horses. Veterinary Microbiology. 2014;**173**(3-4):232-240. DOI: 10.1016/j.vetmic.2014.07.015

[69] van Rooij EM, Rijsewijk FA, Moonen-Leusen HW, Bianchi AT, Rziha HJ. Comparison of different prime-boost regimes with DNA and recombinant Orf virus based vaccines expressing glycoprotein D of pseudorabies virus in pigs. Vaccine. 2010;**28**(7):1808-1813. DOI: 10.1016/j.vaccine.2009.12.004

[70] Fischer T, Planz O, Stitz L, Rziha HJ. Novel recombinant parapoxvirus vectors induce protective humoral and cellular immunity against lethal herpesvirus challenge infection in mice. Journal of Virology. 2003;**77**(17):9312-9323. DOI: 10.1128/JVI.77.17.9312-9323.2003

[71] Henkel M, Planz O, Fischer T, Stitz L, Rziha HJ. Prevention of virus persistence and protection against immunopathology after Borna disease virus infection of the brain by a novel Orf virus recombinant. Journal of Virology. 2005;**79**(1):314-325. DOI: 10.1128/JVI.79.1.314-325.2005

[72] Voigt H, Merant C, Wienhold D, Braun A, Hutet E, Le Potier MF, et al. Efficient priming against classical swine fever with a safe glycoprotein E2 expressing Orf virus recombinant (ORFV VrV-E2). Vaccine. 2007;**25**(31):5915-5926. DOI: 10.1016/j.vaccine.2007.05.035

[73] Rohde J, Amann R, Rziha HJ. New Orf virus (Parapoxvirus) recombinant expressing H5 hemagglutinin protects mice against H5N1 and H1N1 influenza A virus. PLoS One. 2013;**8**(12):e83802. DOI: 10.1371/journal.pone.0083802

[74] Rohde J, Schirrmeier H, Granzow H, Rziha HJ. A new recombinant Orf virus (ORFV, Parapoxvirus) protects rabbits against lethal infection with rabbit hemorrhagic disease virus (RHDV). Vaccine. 2011;**29**(49):9256-9264. DOI: 10.1016/j.vaccine.2011.09.121

[75] Fiebig HH, Siegling A, Volk HD, et al. Inactivated orf virus (Parapoxvirus ovis) induces antitumoral activity in transplantable tumor models. Anticancer Research. 2011;**31**(12):4185-4190

[76] Rintoul JL, Lemay CG, Tai LH, Stanford MM, Falls TJ, de Souza CT, et al. ORFV: A novel oncolytic and immune stimulating parapoxvirus therapeutic. Molecular Therapy. 2012;**20**(6):1148-1157. DOI: 10.1038/mt.2011.301

[77] Tai LH, de Souza CT, Belanger S, Ly L, Alkayyal AA, Zhang J, et al. Preventing postoperative metastatic disease by inhibiting surgery-induced dysfunction in natural killer cells. Cancer Research. 2013;**73**(1):97-107. DOI: 10.1158/0008-5472.CAN-12-1993 80

[78] Choi AH, O'Leary MP, Chaurasiya S, Lu J, Kim SI, Fong Y, et al. Novel chimeric parapoxvirus CF189 as an oncolytic immunotherapy in triple-negative breast cancer. Surgery. 2018;**163**(2):336-342. DOI: 10.1016/j.surg.2017.09.030

[79] Spyrou V, Valiakos G. Orf virus infection in sheep or goats. Veterinary Microbiology. 2015;**181**(1-2):178-182. DOI: 10.1016/j.vetmic.2015.08.010

Section 4

Applications of Systems Biology

Chapter 5

Application of Genomic Data for PCR Screening of Bet v 1 Conserved Sequence in Clinically Relevant Plant Species

Jana Žiarovská and Lucia Zeleňáková

Abstract

Bet v 1 is a highly immunogenic protein, which is the main cause of sensitivity to birch pollen and is described as the main birch allergen. Despite the structural similarity, Bet v 1 homologs show different properties and immunoreactivity. Here, the bioinformatic algorithms were applied for known Bet v 1 homologous nucleic acids sequences to find homology and conserved regions. Genomic sequences of PR proteins of two different fruit species, which allergens belong to PR proteins of the same type as Bet v 1, were selected to design degenerate primers. Subsequently, screening of the presence of Bet v 1 conserved genomic sequence was performed in 45 clinically relevant plant species.

Keywords: genomic sequences, Bet v 1, conserved region, degenerate primers, PCR screening

1. Introduction

Genomic knowledge about major birch pollen allergen is very well known for quite a long time. In the last 30 years, many of different homologs for Bet v 1 have been cloned, and many of their products were characterized from the allergenic point of view. Molecular profiling of allergic sensitization has helped to elucidate the immunological connections of allergen cross-reactivity, whereas advances in biochemistry have revealed structural and functional aspects of allergenic proteins in the last decades [1]. Bet v 1 has been identified as existing in three subfamilies, based on the sequence similarity. The most precise identification is actually done for major birch allergen Bet v 1 that was firstly identified in *Betulla verrucosa* [2]. Bet v 1 is reported in vascular plants as common ones. The first class, pathogenesis-related protein family (PR-10) is expressionally connected to the pathogen attack or abiotic stress. Highest concentrations of PR-10 proteins were found in reproductive tissues (pollen, seeds and fruits) [3] and were described with a high level of similarity with the human lipocalin 2. Birch Bet v 1 and human lipocalin 2 possess specific structures that allowed them to bind iron. Bet v 1 turns to in the situation when it is not binding iron. This subsequently affects Th2 cells of the human immune system [4]. The ribonuclease activity of PR proteins is known to be activated under the function in antiviral pathway [5]. The other subfamilies of

IntechOpen

Bet v 1 allergens are reported as major latex proteins and ripening-related proteins in the latex of opium poppy [6, 7]. The last one is reported to be proteins containing members with S-norcoclaurine synthase activity and is involved in alkaloid biosynthesis [8].

Bet v 1 belongs to panallergens, specifically to PR-10 proteins. Location of Bet v 1 for IgE recognition is the result of the protein chain composition, coming in close proximity of molecules that are spaced apart from the stretched chain. This conformation is disrupted by heat that is why Bet v 1 is defined as thermolabile and it became nonallergenic by cooking or heat processing of fruit. Different denaturation temperatures of the Bet v 1 allergen exist for different individual isolated homologs and their isoforms. A pH value and other thermodynamic and physicochemical properties have a denaturing effect beside the temperature alone [9]. In general, all the PR-10 proteins are labile proteins when comparing them to most of other food allergens [10]. The naturally occurring Bet in 1 consists of several isoforms with a molecular weight of about 17.5 kDa. These isoforms share a high percentage of the same sequences but may have a very different allergenic potential [11, 12]. There are currently more than 20 isoforms found on the IUIS Allergen Nomenclature subcommittee website (http://www.allergen.org). When regarding a total amount of Bet v 1 in grain pollen, about 35% represents the hyperallergenic isoform of Bet in 1.0101 and this is also described in the literature as Bet v 1a [12, 13]. This isoform plays an important role in the development of allergies. It is characterized as the most allergenic isoform. It is used to produce recombinant proteins [11, 12]. Hypoallergenic isoforms are potential candidates for allergen-specific immunotherapy [14]. Despite the structural similarity, Bet v 1 homologs show different properties and immunoreactivity. An example of this is Bet v 1 l - a hypoallergenic isoform, differing in only nine amino acids from highly allergenic Bet v 1a [15]. It is assumed that these different immunological properties are the result of the changing dynamics and stability of the composition of the molecules [16, 17]. Bet v 1 homology is a typical example of a pollen-food allergy syndrome (**Figure 1**). In this case, homology is very close, epitope matching can be up to 90% [18]. Food hypersensitivity to apple, carrot, hazelnuts or celery has been developed in 50–90% of individuals with pollen allergy due to this cross-reactivity in the moderate environment conditions of Europe [19].

Figure 1.
Clinically relevant Bet v 1 homology.

Bet v 1 homology is referred to as birch-fruit-vegetables-nuts syndrome. The most common are plants from the family *Rosaceae* and *Apiaceae*. Similarities in amino acid sequences were found in different plants and foods [20, 21] but a fruit similarity prevails (**Table 1**). Most often, allergens are located in fruit pulp. With respect to homology to the main birch allergen Bet v 1, it can be noted that in areas where the incidence of birch is not quite typical, for example, in southern Europe, sensitivity to Bet in 1 homologs occurs in trees that are similar to alder, hazel, beech and grass allergens [22]. Pomegranate, edible chestnuts, raspberries, spices may also be mentioned. Hrubiško et al. [23] mentioned the cross-reactivity of birch pollen with walnut, almonds, avocados, cherry, plum, peas or asparagus.

Silver Birch is native in most of Europe, northwest Africa and western Siberia, but absent in the southern parts of Europe. It is the most common tree found in Scandinavia and the Alps and a potent pollen producer in those areas. In all of those areas, birch is the most relevant spring pollen allergen relevant during the period from March to May (**Figure 2**).

Atmospheric concentrations of birch pollen grains were monitored [24] and the matched major birch pollen allergen Bet v 1 simultaneously across Europe. The major birch pollen allergen Bet v 1 was determined with an allergen-specific ELISA. The average European allergen release from birch pollen was 3.2 pg. Bet v 1/pollen and the average allergen release in 2009 did not differ substantially between countries. However, a>10-fold difference between daily allergen releases per pollen was detected in all countries. Results of aeropalynological observations in Kiev were reported [25] to be carried out with a gravimetric method. The most abundant pollen types were as follows: *Betulaceae* (21%), *Chenopodiaceae* (10%), *Ambrosia* (10%), *Artemisia* (9%) *Pinaceae* (8%) and *Poaceae* (6%).

A real-time PCR method based on SYBR GREEN technology was developed to analyze the different Bet v 1 expression level [26]. The expression of Betv1 allergen gene was analyzed upon various growth places around Kiev of tested birch pollen samples. Sample from forest growth condition was chosen as a calibrator for

Plant	Bet v 1 homolog	Protein similarity
Apple	Mal d 1	56–63%
Hazelnut	Cor a 1	67%
Peach	Pru p 1	70–73%
Kiwi	Act d 8	53%
Carrot	Dau c 1	37%
Apricot	Pru ar 1	56%
Cherry	Pru av. 1	59–70%
Pear	Pyr c 1	57%
Peanuts	Ara h 8	46%
Celery	Api g 1	41%
Soy	Gly m 4	45%
Strawberries	Fra a 1	53%
Raspberries	Rub I 1	55%

Table 1.
Protein homology of Bet v 1.

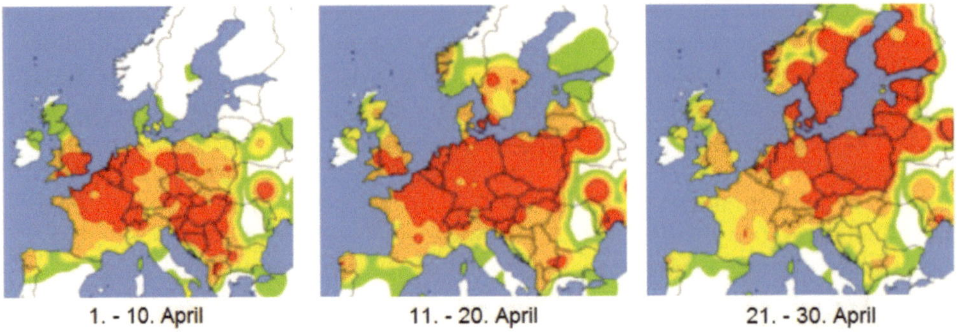

Figure 2.
Pollen load map of Europe during the April (https://www.polleninfo.org). Pollen counts are differed by color: green—low; yellow—mild; orange—high; red—very high.

expression analyses. qRT-PCR showed a variation in the abundance of allergen transcripts among the samples from different places of growth (**Figure 3**). In samples from urbanized area was the expression of Betv1 allergen in average 1.5× higher (ranged from 0.77 up to the 2× higher) when comparing to the forest sample served as a calibrator. In samples from borders of the urbanized area was the expression of Betv1 allergen only 0.55× higher when comparing to the forest sample.

These findings are interesting when comparing them with those findings [24] that reported that extracts from pollen collected in urban areas had higher chemotactic activity on human neutrophils compared to pollen from rural sites, although the allergen content remained unchanged. Questions about the exact correlations between the expression level and allergenic potential need are to be answered in further research.

Actually, different primary genomic data are available for Bet v 1 isoforms originated from birch (**Table 2**) and only limited information exist about its transcriptomic characteristics.

Beside the Bet v 1 – basic allergen component of birch pollen pelvis, minor components exist as well and some of them are clinically relevant too. Allergens of molecular weights of 29.5, 17, 12.5 and 13 kDa had been isolated form birch pollen. The following allergens have been characterized (except of Bet v 1): Bet v 2, a 15 kDa, a profiling; Bet v 3, a 24 kDa calcium-binding protein; Bet v 4, a 9 kDa calcium-binding protein; Bet v 5, a 35 kDa isoflavone reductase-related protein; Bet v 6, a 30–35 kDa protein, phenylcoumaran benzylic ether reductase; Bet v 7, a 18 kDa protein, a cyclophilin and Bet v 11 (www.phadia.com).

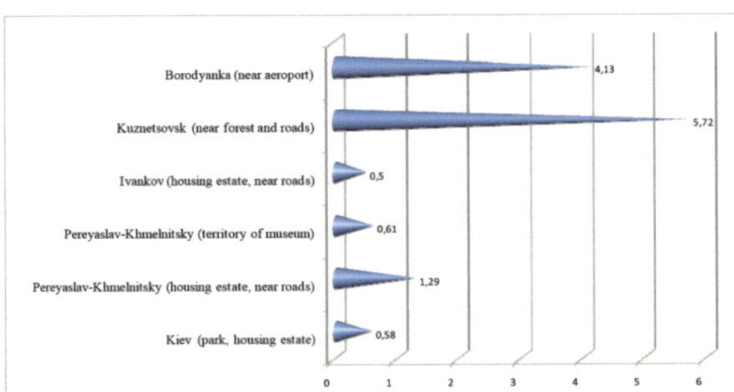

Figure 3.
Expression ratios of Bet v 1 for analyzed birch pollen samples from Ukraine [26].

Bet v 1variant	GenBank accession	Bet v 1variant	GenBank accession
Bet v 1.0101	X15877; Z80098; Z80099	Bet v 1.0115	Z72438.1
Bet v 1.0102	X77266; X77270	Bet v 1.0116	AJ001555.1
Bet v 1.0103	X77267	Bet v 1.0117	AJ006908.1
Bet v 1.0104	X77268; X77274	Bet v 1.0118	AJ006914.1
Bet v 1.0105	X77269	Bet v 1.0119	DQ296603.1
Bet v 1.0106	X77271	Bet v 1.0201	X77200
Bet v 1.0107	X77273	Bet v 1.0202	X77265
Bet v 1.0108	Z80100	Bet v 1.0203	X77272
Bet v 1.0109	Z80101	Bet v 1.0204	X81972; X82028
Bet v 1.0110	Z80102	Bet v 1.0205	Z72431.1
Bet v 1.0111	Z80103	Bet v 1.0206	AJ001556.1
Bet v 1.0112	Z80104	Bet v 1.0207	EU526193.1
Bet v 1.0113	Z80105	Bet v 1.0301	X77601.1
Bet v 1.0114	Z80106	—	—

Table 2.
Available genomic data of birch Bet v 1 isoform stored in public databases.

2. Bet v 1 genomics and *in silico* analysis

From a theoretical point of view, in nature, a protein similarity or analogy to the protein antigen exists to any not only in the plant but also in the animal kingdom. Evidence for this is antigens, particularly those with allergenic potential. From a practical point of view, there is such a similarity for almost every protein and is called homology. Proteins that are similar are referred as the protein family/superfamily. There are a huge number of protein families, many of which have been confirmed to be with allergenic activity [18]. In homology, it is the result of a common evolutionary origin. Homologous genes can be characterized as two or more genes derived from a common original DNA sequence [27]. When identifying genes in the model species and related species, it is often important to distinguish genes mutually linked directly by the species and genes that have been duplicated independently from them. These are two types of homologous genes, orthologs and paralogs with many definitions of them. A status where homology is the result of gene duplication, that is, the two copies remain side by side during the body's past (e.g., alpha and beta hemoglobin), that is why, the genes are called paralogs (para = parallel, analogous, concurrent). In a situation where homology is the result of speciation, that is, the process of generating species, and the past of the gene reflects the past of the species (e.g., alpha hemoglobin in both humans and mice), it should be about orthologs (ortho = exact). Orthologs are genes that are associated with a common origin, genes of different species that have evolved from a common ancestral gene, are called "true" homologs. These genes tend to maintain the same function as the gene they developed from during development process. The identification of orthologs is crucial for a reliable prognosis of gene function in novel genes. Paralogs are genes associated with duplication in the genome. They develop new features even when they are associated with the original function. They deviate from each other within the species. Unlike orthologs, the paralogs gene is a new gene that has a new function. During gene duplication, one copy of the gene is

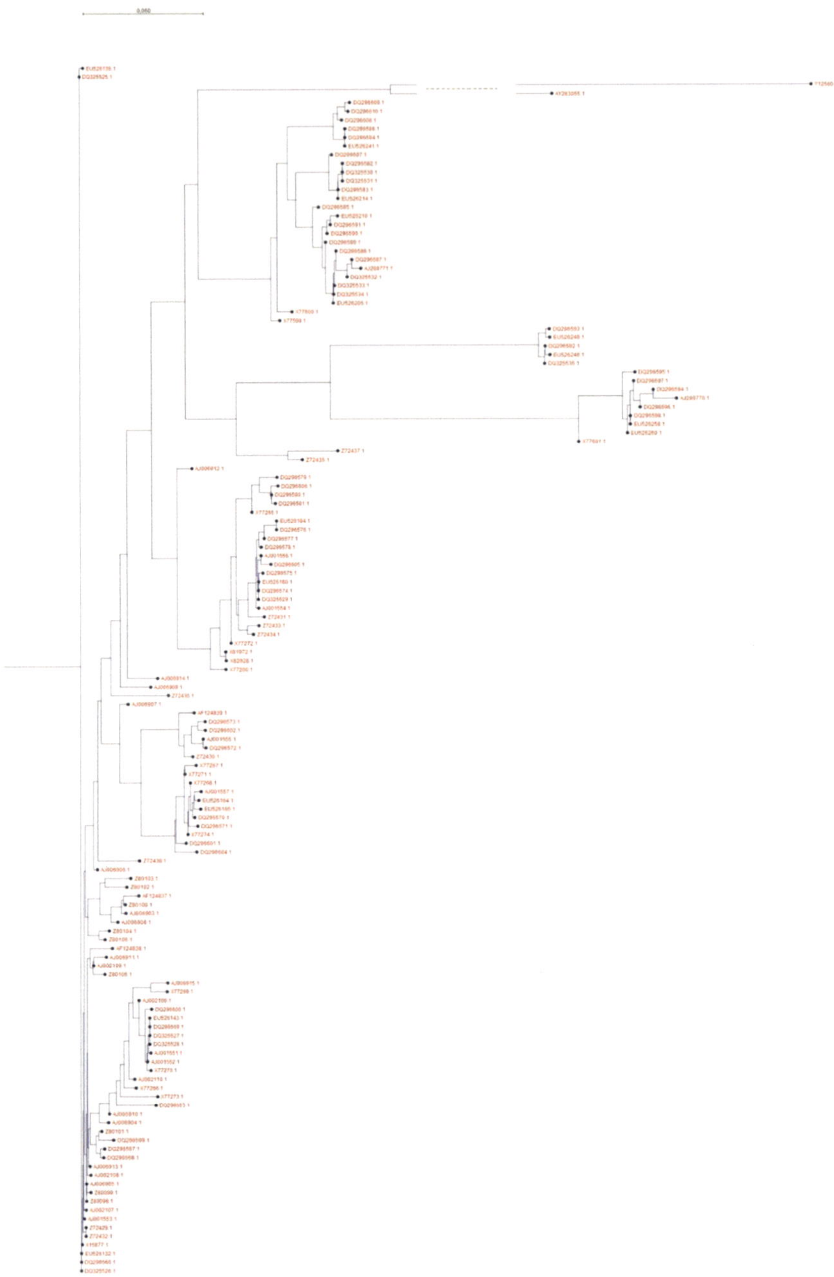

Figure 4.
Sequence phylogenesis of Bet v 1 isoallergens.

mutated to produce a new gene with a new function, although the function often relates to the role of the generic gene [27–29]. Paralogs may result from different types of gene duplication, unequal crossing-over, transposon-mediated duplication or polyploidy, that is, increase in the number of chromosomes in the cell nucleus above the normal diploid state [29]. In the case of the molecular systematics of organisms, it is desirable that the studied sequences are homologously specific - they are called orthologous [30].

Orthologs exist in genomes in a single copy that performs the same function in all organisms examined. The series of evolutionarily conserved genes are paralogs during the evolution; they were done with one or more duplications, followed by the separation of the structure and the functions to the loss of some copies. In some organisms (e.g., in higher plants), the determination of orthologs and paralogs is problematic, their genomes have undergone a series of gene duplications and loss of individual copies of genes. Gene duplication is understood as the source of new genes with new features, but it is not always a fundamental transformation of gene function. Duplicated genes often retain a certain degree of functional overlap that is in certain conditions can be manifested as redundancy.

Different Bet v 1 isoforms are relevant as to be naturally existed—a, b, c, d, e, f, g, j and l. When describing the process of the induction of type I allergy, they differ in reaction mechanism with the IgE from patients, and it is reported in [7] that comparison of *in vitro* and *in vivo* IgE binding activity is influenced by the six amino acid residues at different positions of the Bet v 1 molecule. *Betula verrucosa* (*pendula*) Bet v 1 is well known on the nucleotide level, too and 47 isoallergens sequences are stored in the NCBI database for its mRNA with the different level of their sequence identities. Dendrogram of phylogenetic similarity of the Bet v 1 isoallergens sequences with the gene coding Betv1 (X15877.1) is illustrated in **Figure 4**.

3. Technical approaches and methodologies for PCR screening of Bet v 1 sequences in plants

Bioinformatics provides an interdisciplinary tool, that is used to manage and analyze biological data and known sequences of nucleic acids. Many features of nucleic acids can be used in bioinformatic algorithms as motifs for description of their genomic variability and their better understanding. Individual sequence motifs are recognized by their order and nucleotide preference, and many motif discovery algorithms have been used in different molecular or bioinformatic studies [31–34].

Here, the bioinformatic algorithms were applied for known Bet v 1 homologous sequences what makes them suitable for applying bioinformatic tools such as BLAST [35] to find homology or conserved regions. The first step was to align the individual isoforms and their variants with each other. First, isoforms that exist in two or three variants in the database were compared to each other, namely Bet v 1.0101, Bet v 1.0102, Bet at 1.0104 and Bet at 1.0204. **Table 3** shows results of the sequence alignment of the variants of the Bet isoform at 1.0101. All three isoforms are linear mRNAs, differing only in the number of base pairs. Records Z80099.1 and Z80098.1 have the same number of base pairs. Their overlaps and query cover are up to 100% and the identity 99%.

Bet v 1.0102 can be found in the NCBI under the names of *B. verrucosa* Bet v 1d mRNA (mRNA linear and 677 bp) and *B. verrucosa* Bet in 1 h mRNA (also linear mRNA with 677 bp). They possess a 100% query cover and 99% identity using when using the megablast algorithm. Similar, 100% query cover and 99% identity exist for of the Bet v 1.0104 (*B. verrucosa* Bet v 1f mRNA and *B. verrucosa* Bet in 1i mRNA, both with 572 base pairs). Both searches for Bet v 1.0204 in NCBI are mRNA linear, *B. verrucosa* mRNA for the Bet v 1 m isoform has 687 bp, unlike *B. verrucosa* mRNA for Bet v1n, isoform of birch pollen allergen with 737 bp. Their overlap is 91% with 99% match. After a previous comparison, the consistency of the individual sequences can be as very high, so the variants of the isoforms with the highest number of base pairs were used in the next part of the biological analysis.

Using the BLAST algorithm, individual isoforms corresponding to genomic DNA or mRNA sequences were aligned to each other. The following isoforms are DNA sequences: Bet v 1.0115, Bet v 1.0116, Bet at 1.0119, Bet at 1.0205, Bet v 1.0206, Bet at 1.0207. These accessions have a different number of base pairs. An exception from the point of view of the source exist - Bet v 1.0207 (EU526193.1), with the

Name, type of nucleic acid, number of nt	Accession	Query cover		Identity	
Birch mRNA for pollen allergen BetvI, mRNA linear, 691 bp	X15877.1	Z80098.1	Z80099.1	Z80098.1	Z80099.1
		69%	69%	99%	99%
B. verrucosa mRNA for pollen allergen Betv1 (clone 224), mRNA linear, 483 bp	Z80098.1	Z80099.1		Z80099.1	
		100%		99%	
B. verrucosa mRNA for pollen allergen Betv1 (clone 2230), mRNA linear, 483 bp	Z80099.1	x		x	

Table 3.
Alignment of Bet v 1.0101 nucleotide isoforms that are stored in public databases.

source organism *Betula lenta*, bust stiff. The rest of the aligned sequences have the source organism *Betula pendula* (syn. *B. verrucosa*). As isoforms of one allergen, they are very similar to each other (**Table 4**).

Number of nucleotide differences among Bet v 1 isoforms for the conserved part based on the NCBI data are summarized in **Figure 5**.

The aim for the design of degenerate primers and their subsequent application in the analysis is the basic description and molecular classifications of allergens; finding of correlations between sequence and structural similarities and cross-reactivity between homologous allergens. Genomic knowledge of allergens also helps to define their common properties and will be helping to clear possible factors that cause allergenic potential in the future [37]. Basis necessity in degenerate primer designing is an alignment of selected nucleotide sequences [38]. Here, Bet v 1 was used as a model to analyze functionality of degenerate primers in clinically relevant cross species screening of genomic sequences of allergens (**Table 5**).

Bet v 1 is standardly used as a model pollen protein PR-10 allergen in different types of research aims [12]. Genomic sequences of PR proteins of two different fruit species which allergens belong to PR proteins of the same type as Bet v 1 were selected to design

NCBI accession	Query cover % / Identity %					
	Z72438.1	AJ001555.1	DQ296603.1	Z72431.1	AJ001556.1	EU526193.1
Bet v 1.0115 **(Z72438.1)**						100% 91%
Bet v 1.0116 **(AJ001555.1)**	99% **94%**					
Bet v 1.0119 **(Q296603.1)**	94% **95%**	94% **93%**				
Bet v 1.0205 **(Z72431.1)**	100% **92%**	100% **91%**	99% **90%**			
Bet v 1.0206 **(AJ001556.1)**	99% **92%**	100% **91%**	99% **90%**	99% **98%**		
Bet v 1.0207 **(EU526193.1)**	94% **91%**	95% **91%**	99% **90%**	94 **97%**	95% **98%**	

Table 4.
Alignment of described Bet v 1 isoforms.

High rate cross-reactions		Low rate cross-reactions		Supposed cross-reactions	
Species	**Genomic data**	**Species**	**Genomic data**	**Species**	**Genomic data**
Corylus avellana	DNA/mRNA	*Carpinus betulus*	mRNA	*Ulmus* spp.	N/A
—	—	*Fraxinus excelsior*	mRNA	*Artemisia absinthium*	mRNA
—	—	*Fagus sylvatica*	mRNA	*Secale cereale*	mRNA
—	—	*Quercus robur*	N/A	*Triticum aestivum*	N/A
—	—	—	—	*Phleum pratense*	mRNA
—	—	—	—	*Lolium perenne*	mRNA
Malus domestica	DNA/mRNA	*Prunus armeniaca*	DNA/mRNA	*Litchi chinensis*	N/A
Prunus avium	mRNA	*Prunus domestica*	mRNA	*Mangifera indica*	mRNA
Prunus persica	mRNA*	*Pyrus communis*	mRNA	*Citrus sinensis*	mRNA*
Prunus persica v. nucipersica	N/A	*Actinidia chinensis*	mRNA	*Castanea sativa*	DNA/mRNA
—	—	*Musa spp.*	mRNA*	—	—
Apium graveolens	mRNA	—	—	*Capsicum annum*	mRNA*
Daucus carota	mRNA	—	—	*Spinacia oleracea*	mRNA*
Petroselinum crispum	DNA/RNA	—	—	*Pastinaca sativa*	N/A
—	—	—	—	*Brassica napus*	mRNA*
—	—	—	—	*Cucurbita pepo*	mRNA*
Corylus avelana	DNA/mRNA	*Arachis hypogaea*	DNA/mRNA	—	—
Juglans regia	mRNA*	*Prunus dulcis*	DNA/mRNA	—	—
—		*Foeniculum vulgare*	N/A	*Matricaria recutita*	N/A
—	—	*Carum carvi*	N/A	Curry	N/A
—	—	*Pimpinella anisum*	N/A	Pepper	N/A
—	—	*Coriandrum sativum*	N/A	Black pepper	N/A
Solanum tuberosum	mRNA*	—	—	*Glycine max*	mRNA*

High rate cross-reactions		Low rate cross-reactions		Supposed cross-reactions	
Species	**Genomic data**	**Species**	**Genomic data**	**Species**	**Genomic data**
		—	—	*Fagopyrum esculentum*	N/A
		—	—	*Olea europaea*	DNA/mRNA
		—	—	*Hevea brasiliensis*	mRNA*

**Predicted sequences only.*

Table 5.
Clinically high important cross-allergy species to birch pollen and its genomic information availability about PR-10 class allergens.

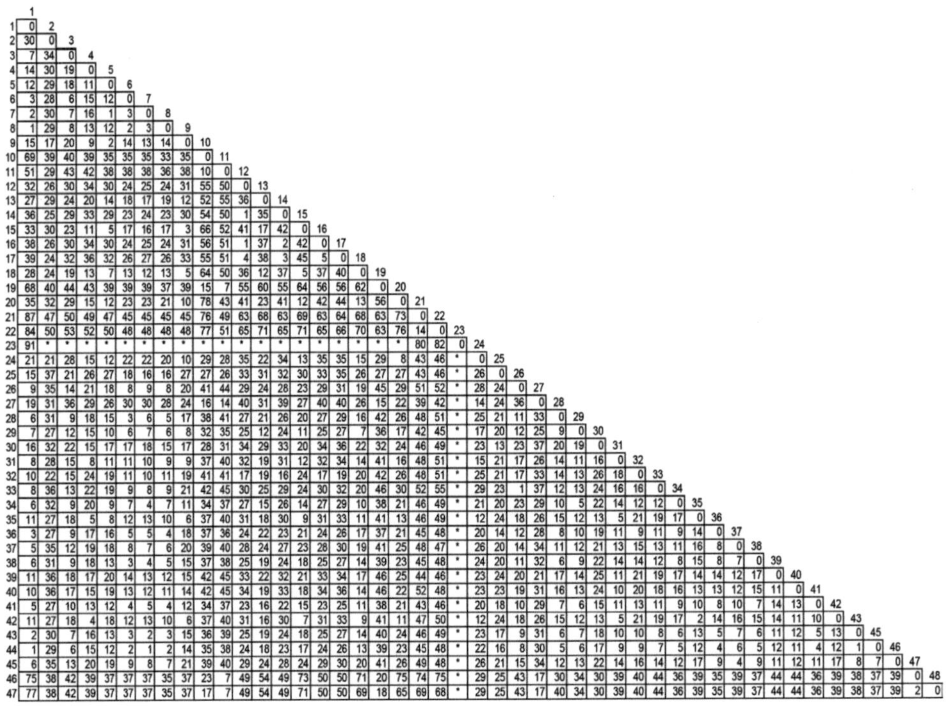

Figure 5.
*Number of nucleotide differences among Bet v 1 isoforms for the conserved part based on the NCBI data. *No sequence homology found; all the isoforms are compared to: X15877.1 and the sequences are coded as follows: 1—AF124839.1; 2—AF124838.1; 3—AF124837.1; 4—AJ002110.1; 5—AJ002109.1; 6—AJ002108.1; 7—AJ002107.1; 8—AJ002106.1; 9—X77200.1; 10—X77272.1; 11—X77274.1; 12—X77273.1; 13—X77271.1; 14—X77270.1; 15—X77268.1; 16—X77267.1; 17—X77266.1; 18—X77265.1; 19—X77269.1; 20—X77599.1; 21—X77600.1; 22—X77601.1; 23—Y12560.1; 24—AJ006915.1; 25—AJ006914.1; 26—AJ006913.1; 27—AJ006912.1; 28—AJ006911.1; 29—AJ006910.1; 30—AJ006909.1; 31—AJ006908.1; 32—AJ006907.1; 33—AJ006905.1; 34—AJ006904.1; 35—AJ006903.1; 36—AJ006906.1; 37—Z80106.1; 38—Z80105.1; 39—Z80103.1; 40—Z80102.1; 41—Z80101.1; 42—Z80100.1; 43—Z80099.1; 44—Z80098.1; 45—Z80104.1;46—X82028.1; 47—X81972.1 [36].*

degenerate primers and to find conserved sequence, that is, the base sequence of the DNA molecule that remained essentially unchanged and thus maintained during the development [39]. *Malus domestica* was used in the selection as a typical fruit caused cross-allergy and *Vitis vinifera* was used in the selection as a species with an allergenic protein homology but without a high clinical relevance. Based on the alignment analysis reported above, Bet v 1 promoter exons (**Table 6**) were used in the *in silico* analysis.

Accession number in NCBI	Description	Type of nucleic acid	Length
AJ289770.1	Betula pendula ypr10b gene, promoter region and exons 1–2.	DNA linear	2687 bp
AJ291705.2	*Vitis vinifera* PR10.1 gene for class 10 pathogenesis-related protein.	DNA linear	1235 bp
AF020542.1	*Malus domestica* major allergen Mal d 1 gene, complete cds.	DNA linear	2253 bp

Table 6.
Genomic sequences used for the conserved region identification.

These sequences were aligned by BLAST as conservative sequences can be identified by homologous searching using this too [39]. Specifically, blastn was used for inter-species comparisons with the result showed in **Figure 6** where alignment of sequences AJ289771.1, AJ291705.2, AF020542.1 to sequence AJ289770.1 can be seen.

The design of degenerate primers for optimal PCR amplification should be based on a conserved region with a length of approximately 200–500 base pairs [38] what is the length that was positively identified in the screened Bet v 1 homologs. Degenerate primer is a mixture of oligonucleotide sequences, each of which has a slightly different sequence, that is, there are several probable bases in it. This extends the range of sequences that can be amplified. This is a sequence of approximately 20–25 bp in length, but the forward and reverse primers must be sufficiently distant from each other, it is another characteristic that was identified positively in the aligned sequences. Based on the obtained results, degenerate primers were designed in this region (**Figure 7**) that provide an amplicon with the length of approximately 365 bp.

Plant material of clinically relevant Bet v 1 (high rate and low rate cross-reactions) cross-reactive plant species and spices were used for the PCR screening analysis. Birch DNA was used as a positive control in the analysis. Total genomic DNA was extracted following the instructions of GeneJET Plant Genomic DNA Purification Mini Kit (Thermo Scientific) or NucleoSpin® Food (Macherey-Nagel). Nanodrop Nanophotometer™ was used for quantity and quality analysis of the extracted DNA. PCR amplifications were performed in a Bio-Rad C1000™ Thermocycler with the following program: an initial denaturation step at 95°C for 5 min followed by

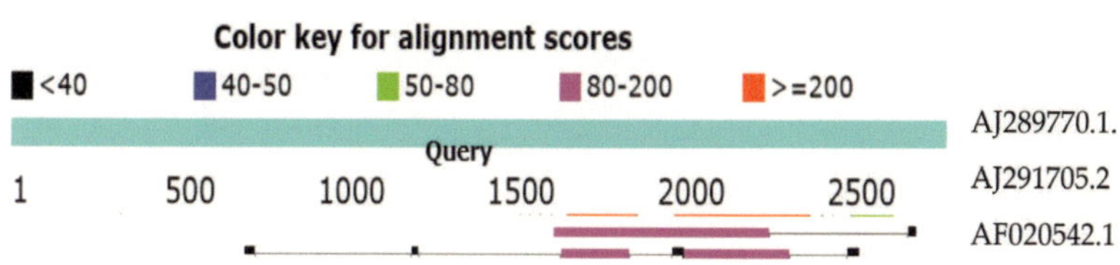

Figure 6.
Most conserved regions of Bet v 1 homologs in apple and grape.

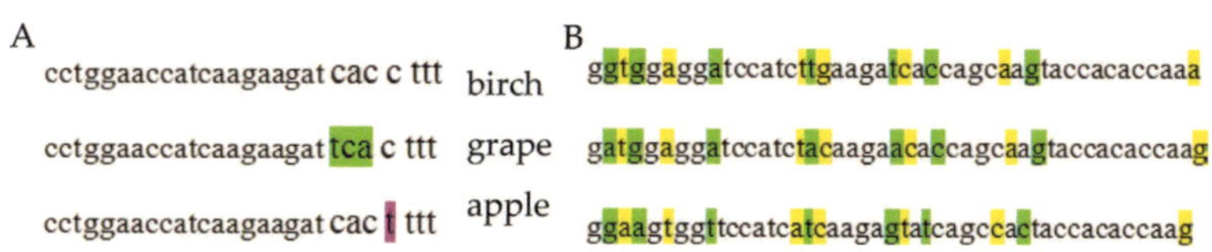

Figure 7.
Selected parts from conservative regions in apple and grape homologs of Bet v 1 that were used for forward (A) and reverse (B) primers designation.

40 cycles at 95°C for 45 s, 54°C for 45 s, and 72°C for 35 s with a final cycle at 72°C for 10 min. The amplified products were inspected by electrophoresis in 1.5% agarose in a 1 × TBE buffer, visualized after GelRed™ staining and photographed under UV light.

Using the degenerate primer pair that was designed on the basis of identified conservative region of Bet v 1 sequences, in all of the screened plant species, PCR was positive with the exception of two samples—curry and black pepper spice (**Table 7**). Here, in the case of curry, only a very weak amplicon is visible in the agarose gel (**Figure 8**), that is why it can be supposed, that a further optimization of degenerate PCR will give a positive result, too. In the case of black pepper spice, using an alternative DNA extraction protocols should be tested further.

Sequence homology search algorithms became commonly used and efficient tools in molecular genetics [39, 40]. Nowadays, a number of different motifs finding algorithms are available and reported them to be impossible to provide a comprehensive report of all of them. Each algorithm has its own advantages and disadvantages. One of the aims of different patterns discovery is finding of specific motifs in nucleotide or protein sequences for the purpose of better understanding of their structure and function [41]

High rate cross-reactions		Low rate cross-reactions		Supposed cross-reactions	
Species	**Bet v 1 PCR**	**Species**	**Bet v 1 PCR**	**Species**	**Bet v 1 PCR**
M. domestica	+	*C. betulus*	+	*Ulmus* spp.	+
P. avium	+	*F. excelsior*	+	*A. absinthium*	+
P. persica	+	*F. sylvatica*	+	*S. cereale*	+
P.p. v. nucipersica	+	*Q. robur*	+	*T. aestivum*	+
A. graveolens	+	*P. armeniaca*	+	*P. pratense*	+
D. carota	+	*P. domestica*	+	*L.perenne*	+
P. crispum	+	*P. communis*	+	*L. chinensis*	N/A
C. avelana	+	*A. chinensis*	+	*M. indica*	+
J. regia	+	*Musa spp.*	+	*C. sinensis*	+
S. tuberosum	+	*A. hypogaea*	+	*C. sativa*	+
		P. dulcis	+	*C. annum*	+
		F. vulgare	+	*S. oleracea*	+
		C. carvi	+	*P. sativa*	+
		P. anisum	+	*B. napus*	+
		C. sativum	+	*C. pepo*	+
				M. recutita	+
				Curry	—
				Pepper	+
				Black pepper	—
				G. max	+
				F. esculentum	+
				O. europaea	+
				H. brasiliensis	N/A

Table 7.
Results of PCR screening of conserved region of Bet v 1 genomic sequence in clinically relevant plant species.

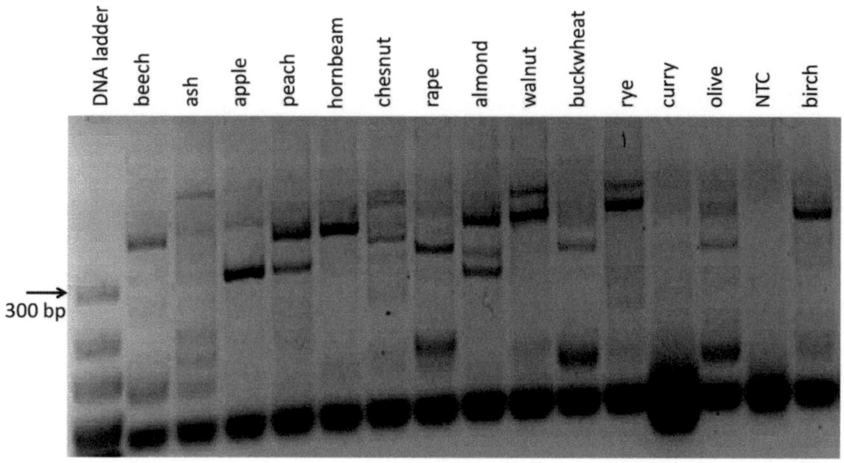

Figure 8.
Electrophoretogram profiles of amplicons of conserved region of Bet v 1 genomic sequence in clinically relevant plant species.

or for their identification [42]. Describing the existing polymorphism is relevant for allergens not only toward its static description, but moreover toward its biological and clinical implication. Different changes in allergen expression are reported for pollution or abiotic stress responses [26, 43, 44]. Very specific knowledge is obtained in the field of the variability of allergenic molecules with respect to the genetic origin of the allergens for different plant species, such as olive, palm date or apple [45–48]. In birch, 13 Bet v 1 putative alleles have been characterized and their occurrence in different cultivars is a matter for future study [49]. Allergens identification has become an integral part of the characteristics of many foodstuffs. The research in this area is important not only from the scientific point of view, but also from the view of impact's to the health as there is an increasing number of people suffering from allergies.

4. Conclusion

A variety of allergens from different fruits were identified based on experimental immunology and molecular biology, that is, by sequencing, leading to gene and protein identification. Whereas allergens are typically described in certain plant species, each of them has a high degree of sequence identity to other proteins in their groups. Among the different fruit allergens, the pathogenesis-related (PR) proteins, classified into 17 families based on sequence, diverse structure, function and biological activity, and they are produced in response to different biotic and abiotic stresses. Allergens of individual plant food sources are very well described and structural details are known as well as the interaction with the immune system of patients. But at the level of regulation and expression of the genes themselves in plants, our knowledge is very limited for the known allergenic proteins. Basic genomic and transcriptomic analysis of the allergens will help to understand their natural genomic background in individual plant varieties and will lead to better personal allergy management in the future.

Acknowledgements

This work was supported by the grants KEGA 007SPU-4/2017 and by The Danube strategy project DS-2016-0051.

Conflict of interest

Authors declare no conflict of interest.

Thanks

The authors would like to thank Ing. Beáta Kováčová for her technical assistance and to colleagues from Department of Biochemistry and Biotechnology, namely Dr. Martin Vivodík for providing DNA of rye and buckwheat for this analysis.

Author details

Jana Žiarovská[1*] and Lucia Zeleňáková[2]

1 Department of Genetics and Plant Breeding, Faculty of Agrobiology and Food Resources, Slovak University of Agriculture in Nitra, Nitra, Slovak Republic

2 Department of Food Hygiene and Safety, Faculty of Biotechnology and Food Sciences, Slovak University of Agriculture in Nitra, Nitra, Slovak Republic

*Address all correspondence to: jana.ziarovska@uniag.sk

References

[1] Asam C, Hofer H, Wolf M, Aglas L, Waallner M. Tree pollen allergens—An update from a molecular perspective. Allergy. 2015;**70**:1201-1211. DOI: 10.1111/all.12696

[2] Liscombe DK, MacLeod BP, Loukanina N. Evidence for the monophyletic evolution of benzylisoquinoline alkaloid biosynthesis in angiosperms. Phytochemistry. 2005;**66**:2501-2520

[3] Markovic-Housley Z, Degano M, Lamba D. Crystal structure of a hypoallergenic isoform of the major birch pollen allergen Bet v 1 and its likely biological function as a plant steroid carrier. Journal of Molecular Biology. 2003;**325**:123-133

[4] Roth-Walter F, Gomez-Casado C, Pacios LF. Bet v 1 from birch pollen is a lipocalin-like protein acting as allergen only when devoid of iron by promoting Th2 lymphocytes. The Journal of Biological Chemistry. 2014;**289**:17416-17421

[5] Park CJ, Kim KJ, Shin R. Pathogenesis-related protein 10 isolated from hot pepper functions as a ribonuclease in an antiviral pathway. Plant Journal. 2004;**37**:186-198

[6] Osmark P, Boyle B, Brisson N. Sequential and structural homology between intracellular pathogenesis-related proteins and a group of latex proteins. Plant Molecular Biology. 1998;**38**:1243-1246

[7] Vieths S, Scheurer S, Ballmer-Weber B. Current understanding of cross-reactivity of food allergens and pollen. Annals of the New York Academy of Sciences. 2002;**964**:47-68

[8] Samanani N, Liscombe DK, Facchini PJ. Molecular cloning and characterization of norcoclaurine synthase, an enzyme catalyzing the first committed step in benzylisoquinoline alkaloid biosynthesis. Plant Journal. 2004;**40**:302-313

[9] Bollen MA, Wichers HJ, Helsper JPFG, Savelkoul HFJ, Boekel MAJS. Thermodynamic characterization of the PR-10 allergens Bet v 1, Api g 1 and Dau c 1 and pH-dependence on Api g 1 and nDau c 1. Food Chemistry. 2010;**119**:241-248. DOI: 10.1016/j.foodchem.2009.06.013

[10] Mills C, Wichers H, Hoffmann-Sommergruber K. The effects of food processing on allergens. In: Mills C, Wichers H, Hoffmann-Sommergruber K, editors. Managing Allergens in Food. Woodhead Publishing; 2007. 336 pp. ISBN: 978-1-84569-028-1

[11] Ferreira F, Hirtenlehner K, Jilek A, Godnik-Cvar J. Dissection of immunoglobulin E and T lymphocyte reactivity of isoforms of the major birch pollen allergen Bet v 1: Potential use of hypoallergenic isoforms for immunotherapy. The Journal of Experimental Medicine. 1996;**183**:599-609 https://www.ncbi.nlm.nih.gov/pubmed/8627171

[12] Ahammer L, Grutsch S, Wallner M, Ferreira F, Tollinger M. NMR resonance assignments of a hypoallergenic isoform of the major birch pollen allergen Bet v 1. Biomolecular NMR Assignments. 2017;**11**:231-234. DOI: 10.1007/s12104-017-9754-7

[13] Wagner S, Radauer C, Bublin M, Hoffmann-Sommergruber K, Kopp T, et al. Naturally occurring hypoallergenic Bet v 1 isoforms fail to induce IgE responses in individuals with birch pollen allergy. Journal of Allergy and Clinical Immunology. 2008;**121**:246-252. DOI: 10.1016/j.jaci.2007.08.006

[14] Valenta R, Campana R, Focke-Tejkl M, Niederberger V. Vaccine development for allergen-specific immunotherapy based on recombinant allergens and synthetic allergen peptides: Lessons from the past and novel mechanisms of action for the future. Journal of Allergy and Clinical Immunology. 2016;**137**:351-357. DOI: 10.1016/j.jaci.2015.12.1299

[15] Bollen MA, Garcia J, Cordewener HGJ, Wichers JH, Helsper PFGJ, Savelkoul FJH, van Boekel AJSM. Purification and characterization of natural Bet v 1 from birch pollen and related allergens from carrot and celery. Molecular Nutrition & Food Research. 2007;**51**:1527-1536. DOI: 10.1002/mnfr.200700119

[16] Freier R, Dall E, Brandstetter H. Protease recognition sites in Bet v 1a are cryptic, explaining its slow processing relevant to its allergenicity. Scientific Reports. 2015;**5**:1-9. DOI: 10.1038/srep12707

[17] Machado Y, Freier R, Scheiblhofer S, Thalhamer T, Mayr M, et al. Fold stability during endolysosomal acidification is a key factor for allergenicity and immunogenicity of the major birch pollen allergen. Journal of Allergy and Clinical Immunology. 2016;**137**:1525-1534. DOI: 10.1016/J.JACI.2015.09.026

[18] Fuchs et al. Potravinová alergie a intolerance. Praha: Mladá fronta; 2016. 448 pp. ISBN 978-80-204-3757-0

[19] Kateris CH. Food allergy and oral allergy or pollen-food syndrome. Current opinion in allergy and clinical immunology. 2010;7:55-60. DOI: 10.1097/ACI.0b013e32833973fb

[20] Subbarayal B, Geroldinger-Simic M, Bohle B. Birch pollen-related food allergy: An excellent disease model to understand the relevance of immunological cross-reactivity for allergy. In: Pereira C, editor. Allergic Diseases—Highlights in the Clinic, Mechanisms and Treatment. IntechOpen; 2012. pp. 141-164. ISBN: 978-953-51-0227-4

[21] Špičák V. Genetika alergie. In: Petru V et al., editors. Dětská alergologie. Praha: Mladá fronta; 2012, 531. ISBN: 9788020425843

[22] Bohle B. The impact of pollen - related food allergens on pollen allergy. Allergy. 2007;**62**:3-10. DOI: 10.1111/j.1398-9995.2006.01258.x

[23] Hrubiško M et al. Alergológia. Osveta: Martin; 2003. 518 p. ISBN: 8080631107

[24] Buters JTM, Weichenmeier I, Ochs S, Pusch G, Kreyling W, Boere AJF, Schober W, Behrendt H. The allergen Bet v 1 in fractions of ambient air deviates from birch pollen counts. Allergy. 2010;**65**:850-858

[25] Savitsky VD, Bezus'ko LG, Butich NG, Tsymbaliuk ZM, Savitska OV, Bezus'ko TV. Airborne pollen in Kiev (Ukraine): gravimetric sampling. Aerobiologia. 1996;**12**:209-211

[26] Žiarovská J, Labajová M, Ražná K, Bežo M, Štefúnová V, Shevtsová T, Garkava K, Brindza J. Changes in expression of Bet V1 allergen of silver birch pollen in urbanized area of Ukraine. Journal of Environmental Science and Health. 2013;**48**:1479-1484

[27] Sheldon E. The Difference Between Orthologous & Paralogous Genes. https://sciencing.com/difference-between-orthologous-paralogous-genes-18612.html

[28] Fitch MW. Distinguishing homologous from analogous proteins. Systematic Biology. 1970;**19**:99-113. DOI: 10.2307/2412448

[29] Cannon SB, Young ND. OrthoParaMap: Distinguishing orthologs from paralogs by integrating comparative genome data and gene phylogenies. BMC Bioinformatics. 2003;**4**:35. DOI: 10.1186/1471-2105-4-35

[30] Cvrčková F. Úvod do praktické bioinformatiky. Praha: Nakladatelství Academia; 2006. 148 pp. ISBN: 80-200-1360-1

[31] Aravind L, Koonin EV. Gleanin non-trivial structural, functional and evolutionary information about proteins by interactive database searches. Journal of Molecular Biology. 1999;**287**:1023-1040. DOI: 10.1006/jmbi.1999.2653

[32] Hertz GZ, Stormo GD. Identifying DNA and protein patterns with statistically significant alignments of multiple sequences. Bioinformatics. 1999;**15**:563-577. DOI: 10.1093/bioinformatics/15.7.563

[33] La D, Livesay DR. Predicting functional sites with an automated algorithm suitable for heterogeneous datasets. BMC Bioinformatics. 2005;**6**:116. DOI: 10.1186/1471-2105-6-116

[34] Rasouli H, Kahrizi D, Ghadernia P. Identification of conserved domains and motifs for *TaWdhn13* gene in *Triticum aestivum* by in silico analysis. Advances in Environmental Biology. 2013;7:586-590

[35] Altschul SF, Madden TL, Schäffer AA, Zhang J, Zhang Z, Miller W, Lipman DJ. Gapped BLAST and PSI-BLAST: A new generation of protein database search programs. Nucleic Acids Research. 1997;**25**:3389-3402. DOI: 10.1093/nar/25.17.3389

[36] Žiarovská J, Zeleňáková L. Central and Eastern European spring pollen allergens and their expression analysis-state of the art. Diversity. 2016;**8**:1-11

[37] Radauer CH, Breitener H. Structure, Allergenicity, and Cross-Reactivity of Plant Allergens. New York: Springer; 2009. pp. 127-151. ISBN: 978-0-387-79208-8

[38] Iserte JA, Stephan BI, Goni SE, Borio CS, Ghiringhelli PD, Lozano ME. Family-specific degenerate primer design: A tool to design consensus degenerated oligonucleotides. Biotechnology Research International. Vol. 2013:38364. DOI: 10.1155/2013/383646

[39] Nawrocki EP, Eddy SR. Infernal 1.1: 100-fold faster RNA homology searches. Bioinformatics. 2013;**29**:2933-2935. DOI: 10.1093/bioinformatics/btt509

[40] Karpov PA, Nadezhdina ES, Yemets AI, Matusov VG, Nyporko AY, Shashina NY, Blume YB. Bioinformatic search of plant microtubule-and cell cycle related serine-threonine protein kinases. BMC Genomics. 2010;**11**:S14. DOI: 10.1186/1471-2164-11-S1-S14

[41] Bailey TL. Discovering sequence motifs. Methods in Molecular Biology. 2008;**452**:231-251. DOI: 10.1007/978-1-60327-159-2_12

[42] Khuri S, Bakker FT, Dunwell JM. Phylogeny, function, and evolution of the cupins, a structurally conserved, functionally diverse superfamily of proteins. Molecular Biology and Evolution. 2001;**18**:593-605. DOI: 10.1093/oxfordjournals.molbev.a003840

[43] Askari H, Edqvist J, Hajheidari M, Kafi M, Salekdeh GH. Effects of salinity levels on proteome of *Suaeda aegyptiaca* leaves. Proteomics. 2006;**6**:2542-2554

[44] Ražná K, Bežo M, Nikolaieva N, Garkava K, Brindza J, Žiarovská J. Variability of *Corylus avellana*, L. Cor A and profilin pollen allergens expression. Journal of Environmental Science and Health. 2014;**49**:639-645

[45] Alché JD, Castro AJ, Jiménez-López JC, Morales S, Zafra A, Hamman-Khalifa AM, Rodríguez-García MI. Differential characteristics of olive pollen from different cultivars: Biological and clinical implications. Journal of Investigational Allergology and Clinical Immunology. 2007;**17**(Suppl 1):17-23

[46] Kwaasi AA, Parhar RS, Tipirneni P, Harfi HA, Al-Sedairy ST. Cultivar-specifi c epitopes in date palm (*Phoenix dactylifera* L.) pollenosis. Differential antigenic and allergenic properties of pollen from ten cultivars. International Archives of Allergy and Immunology. 1994;**104**(3):281-290

[47] Kwaasi AA, Harfi HA, Parhar RS, Collison KS, Al-Sedairy ST, Al-Mohanna FA. Cultivar-specific IgE-epitopes in date (*Phoenix dactylifera* L.) fruit allergy. Correlation of skin test reactivity and IgE-binding properties in selecting date cultivars for allergen standardization. International Archives of Allergy and Immunology. 2000;**123**(2):137-144

[48] Hsieh LS, Moos M Jr, Lin Y. Characterization of apple 18 and 31 kd allergens by microsequencing and evaluation of their content during storage and ripening. Journal of Allergy and Clinical Immunology. 1995;**96**:960-970

[49] Schenk MF, Gilissen LJ, Esselink GD, Smulders MJ. Seven different genes encode a diverse mixture of isoforms of bet v 1, the major birch pollen allergen. BMC Genomics. 2006;7:168

Chapter 6

Insight into the Mechanism of Red Alga Reproduction. What Else Is Beyond Cystocarps Development?

Pilar Garcia-Jimenez and Rafael R. Robaina

Abstract

Volatile growth regulators play an important role in triggering aspects related to red seaweed reproduction. The last 10 years have brought clarification to how ethylene and methyl jasmonate work. Taking two reproductive stages of thalli of red seaweed—fertilised and fertile thalli—as benchmarks and a precise characterisation of the elicitation and disclosure periods of cystocarps, monitoring different gene expressions, namely candidate gene for reproduction and genes encoding proteins involved in biosynthesis pathways of both volatiles and reactive oxygen species, has enabled us to discern the differential behaviour of genes. These studies have also revealed that the volatile-mediated signal could affect cell wall loosening. All in all, studies have shown evidence of putative signalling pathways where volatile signal regulators form part of them at several levels, ranging from disclosure, development to the maturing of cystocarps. This signal information is crucial to determine the final response. The chapter also discusses whether signal transduction is related to different sensing for each volatile and whether this could be elicited in accordance with signal strength. This chapter compiles our current understanding of molecular mechanisms of algal reproduction and how volatile-mediated signals affect other developmental processes.

Keywords: ethylene, genes, methyl jasmonate, red seaweed, volatile growth regulators

1. Introduction

Carposporogenesis in red algae requires the disclosure and development of reproductive structures named cystocarps and cell wall weakness and also requires these reproductive structures to mature. Disclosure of cystocarps, in other words, the period in which the first cystocarps become visible, is elicited by external signals such as volatile growth regulators. On the other hand, controlling the elicitation period is essential for the proper development of cystocarps. If this period does not lead to the disclosure of cystocarps, cell wall loosening will not occur, and these structures will not mature either.

Once the elicitation period occurs, cystocarp development begins with the weakening and relaxation of the cell wall. In Floridophyceae, the accepted view is that the cell wall is made of well-organised layers, whilst the intracellular matrix is comprised of less organised material. Some of components of the cell wall and matrix are sulphated galactans, which have a physiological significance that varies according to the different life stages of the macroalgae [1]. Despite the relationship between the loosening and weakening of the cell wall and the different life stages, the biochemical and molecular mechanisms have not been fully discovered. Evidence suggests that reactive oxygen species, under growth regulator control, are able to cleave cell wall polysaccharide, causing the wall to loosen [2, 3] during reproductive events in seaweed.

The maturity stages of reproductive structures in red algae are complex processes (**Figure 1**), highly co-ordinated and, to a large extent, quite difficult to determine. Unlike some seaweeds where different stages of development of cystocarps are evident and can be recognised [4], in others, the maturity stages are assumed to occur from the beginning of the discloser of the reproductive structures to thalli. In these cases, the maturity process differentiates between two kinds of thalli, the fertilised thalli and fertile thalli. The fertilised thalli are the ones that have both non-visible cystocarps and incipient visible cystocarps. Meanwhile, fertile thalli range from thalli with well-developed cystocarps to those that have fully mature cystocarps (**Figure 2A**).

Changes related to the maturing of thalli are favoured by volatile growth regulators, which also lead to both cystocarps dehiscence with a marked reduction of the maturity period, and the presence of different reproductive structures in the same individual [5]. Moreover, other evidence such as sudden losses of seaweed mats and alternating life cycles could also give cues on how volatile compounds act as a signal to trigger the reproductive process. Actually, seaweeds have a defined reproductive period and are able to discern between volatile signals. The latter leads to the presence of 'putative' volatile receptors although they are not yet known and only a proposed ethylene receptor in red algae has been reported [6].

With this scenario, advances in gene studies have been made by combining different approaches—based on evidence of in vitro culture in the presence of volatile growth regulators and on algal physiology—and thus to decipher the network of interactions between different metabolic pathways that lead the transition from fertilised to fertile thalli. This path can lead to an understanding of a complex network of interacting genes and signal pathways that occurs. Hence one of the key questions is also to unveil how this process can be co-ordinated to work efficiently.

In recent years, great strides have been taken to gain understanding of molecular events in red seaweeds. These endeavours have allowed for a better understanding of the changes that occur during the transition from disclosure to the maturity of

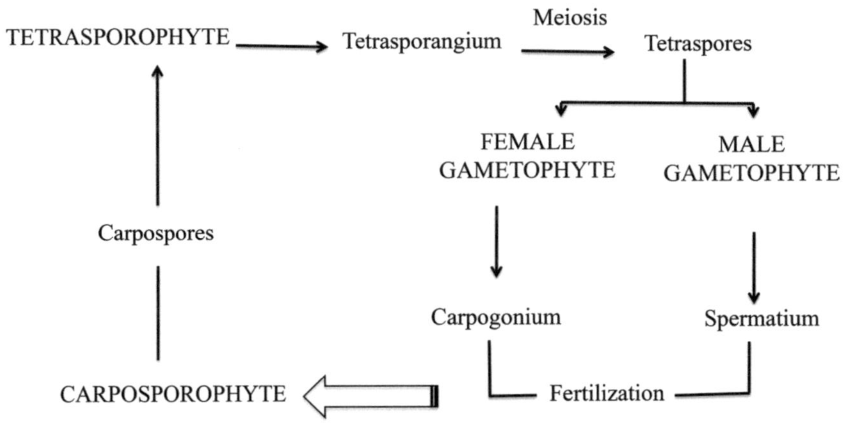

Figure 1.
Diagram of a tri-genetic life cycle in the red alga Grateloupia imbricata comprising the gametophyes (haploids), called carposporophyte, that develops on the female gametophyte after fertilisation, and the sporophyte (diploid).

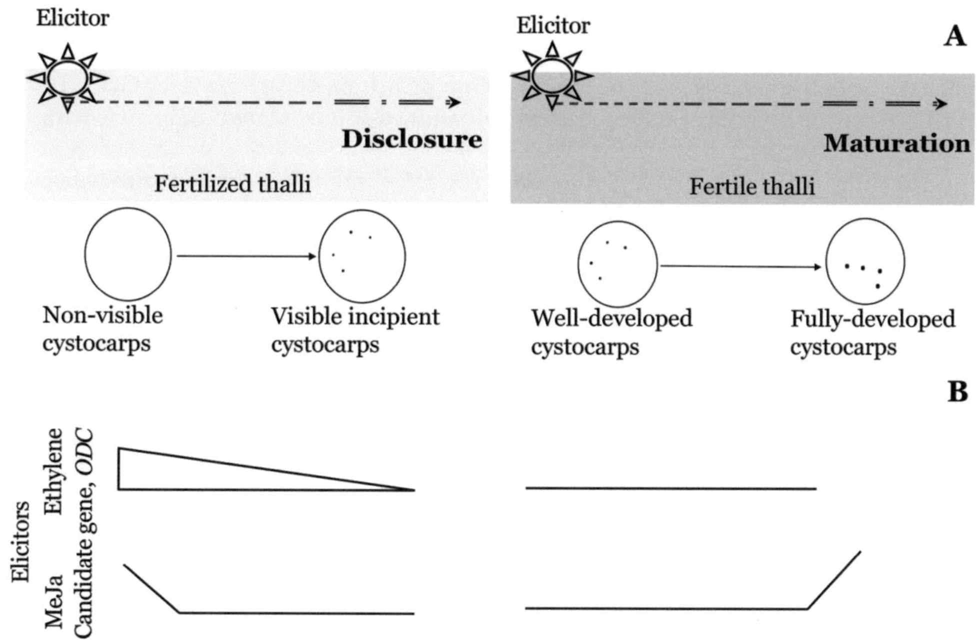

Figure 2.
(A) Schematic showing the timeline for the periods of disclosure and maturity of cystocarps and the corresponding fertilised thalli within cystocarps, ranging from non-visible to visible incipient and fertile thalli from well-developed to fully developed cystocarps. (B) Timeline of gene expression of ODC for fertilised and fertile thalli of Grateloupia imbricata after ethylene and methyl jasmonate treatment. The sloping lines represent significant increase or decrease in gene expression with respect to absolute values (copies μl^{-1}). Horizontal line indicates no changes in gene expression relative to expression in control thalli. MeJa, methyl jasmonate.

cystocarps in response to growth regulator. In this chapter we present our research output in the carragenophytic red seaweed model *Grateloupia imbricata*, compiling our current understanding of molecular mechanisms of algal reproduction and how a volatile-mediated signal can affect other developmental processes. This work does not forget to review other articles, but it does focus on what the state of the art is concerning red seaweed reproduction based on (1) candidate genes, (2) genes that encode cell wall weakness and reactive oxygen species and (3) genes that encode biosynthesis of volatile growth regulators such as ethylene and methyl jasmonate.

2. Candidate gene of reproduction

Growth, development and reproduction of multicellular organisms require precise and multifunctional cell–cell communication events. This is even more necessary in marine seaweed, where changes in irradiation, salinity and temperature, due to the tidal period, affect sporulation and mean that these organisms have to handle and adapt environmental signals in an extremely precise manner to survive. Taking this into consideration, it is easy to understand that algae make quick acclimation—reversible acclimation—and adaptation to the marine environment and that the control of some of these vegetative and reproductive processes is particularly based on short-range signalling.

With this complex net of intervening factors, in order to interpret what is occurring with a particular event, the election of a candidate gene, which represents the manifestation of a trait such as the development and maturity of cystocarps, has provided insights into the carposporogenesis of red seaweeds.

Unlike unspecific genes that are overexpressed under a given condition and are assumed to be responsible for a particular event/trait/action [7–9], our candidate gene encodes the synthesis of the main protein ornithine decarboxylase (ODC, EC 4.1.1.17) responsible for the synthesis of polyamines and is related to the maturing process of cystocarps in seaweeds [10–12].

The differential behaviour of this gene (*GiODC*) and its integration with volatiles contribute to it being chosen as a candidate gene for several reasons. First, the inhibition of ODC enzyme synthesis by the inhibitor DL-α-difluoromethylornithine (DFMO) implies the lowest levels of polyamines. This inhibition also halts the maturity of cystocarps and the eventual release of spores from cystocarps [11]. Second, the enzyme activity of ODC is related to the endogenous levels of polyamines. The reduction in ODC enzyme activity and polyamine levels are also related to the presence of cystocarps [10, 12]. Third, reactive oxygen species are released through polyamine catabolism pathways and are under the control of ODC. During cystocarp development, spermine is accumulated, favouring the process of development and maturity of the reproductive structure. When it exerts an inductive effect, polyamine oxidase enzyme activity increases as the spermine degrades [12]. Fourth, *GiODC* is expressed differently in both the fertile thalli (with cystocarps), than in infertile thalli (vegetative thalli), and in the apical part of fertile thalli, against the basal part of these thalli, as reported using -time quantitative PCR and in situ hybridisation techniques [13]. Fifth, sequencing the upstream region of *GiODC* revealed transcription factors involved in regulation by jasmonate (Myc2, Myc3 and Myc4) and ethylene (RAV, SMZ and Abi4). This means that there is a relationship between volatiles and *ODC* expression [14]. Sixth, monitoring *GiODC* gene expression after treatment with volatiles during the well-defined periods of elicitation and disclosure of cystocarps reveals differential behaviour of this gene, depending on the development and maturity of the cystocarps [14].

Putting all the data together suggests two important conclusions regarding the candidate gene: Expression is dependent on the existence of cystocarps and the kind of growth regulator used to elicit reproduction. Generally, down-expression of the gene candidate goes hand in hand with the presence of cystocarps and points to a quick transduction signal (**Figure 2B**) [15, 16]. Nonetheless, it is worth mentioning that there are two different gene expression patterns that occur when methyl jasmonate is used as an elicitor. Hence, in thalli without visible cystocarps, gene expression is upregulated due to methyl jasmonate signalling (fertilised thalli, **Figure 2B**). Moreover, in thalli containing fully mature cystocarps, other up-expressions are related to the stage of maturity of the cystocarp due to methyl jasmonate (fertile thalli, **Figure 2B**).

Far from being a mismatch for a candidate gene, it is understood that different signals are executed over the course of cystocarp development, and hence one can infer that thalli are able to discern between volatiles; they sense them in order to provide co-ordinated responses [14–16].

3. Genes encoding proteins related to oxidative stress and softening of thalli

In most organisms, factors including drastic changes in temperature, irradiation and desiccation are stressful and potentially destructive. Nonetheless reproduction in algae is also highly regulated by temperature and tidal periods, which has an impact on processes such as sporulation. The generation of reactive oxygen species in turn is triggered by these environmental factors, as can be expected. To ameliorate this situation, organisms display various physiological responses which are often being associated with an increase in the production of proteins that scavenge free radicals and reactive oxygen species (ROSs) [17, 18]. Unlike what has been well studied in higher plants, where stress proteins can be synthesised as a key survival strategy, we know that similar processes can occur, but it remains unclear whether stress proteins are metabolically biosynthesised or whether free radicals can be eliminated by chemical scavenging. Consideration also has to be given to the fact that certain red algae render methylate halides using methyl-transferases that use S-adenosyl methionine (SAM, pivotal compound for the synthesis of ethylene and methyl jasmonate) as the methyl donor. Methylation of halides is a mechanism eliminating halide and sulphide ions, both of which are known to be phototoxic [19, 20].

Beyond this, seaweeds develop strategies to signal events related to growth and development, including the biosynthesis of volatiles. These volatile signals appear to crosstalk with other growth regulators such as polyamines [21]. As an example, polyamines, ethylene and methyl jasmonate share the same precursor—SAM—for these biosynthesis routes. Moreover, ROSs can be also released through metabolic pathways of growth regulators. The contribution of these signal pathways to growth and development is difficult to appraise as volatiles can have synergistic effects on one or more of the other pathways involved in seaweed reproduction, and this combination of all the pathways might give rise to several responses. Ethylene and methyl jasmonate provoke changes in the oxidation state of intermediates during synthesis. These include jasmonates, which are compounds, resulting from lipid oxidation of the cell membrane. In particular, methyl jasmonate is derived from linolenic acid, via lipoxygenase, in which the synthesis of methyl jasmonate activates the oxidative metabolism of polyunsaturated fatty acids, generating ROSs (in the form of O_2, H_2O_2 or OH^-) and oxidised derivatives of polyunsaturated fatty acids [22, 23]. Oxygenated volatile compounds have been shown to not necessarily involve photodamage of cell membranes. Meanwhile the reactivity of the ethylene double bond allows this olefin to be easily converted into a range of intermediates [24].

With this framework, ROSs also have the potential to interact with many cell components and can give rise to several physiological responses, such as when ROS acts as an important signal transduction molecule during growth [18]. Indeed, it has been inferred that ROSs play an important role in softening of thalli and therefore in the development of cystocarps in red seaweed. This is significantly important with the heat shock protein WD40 and cytochrome P450 which are responsible for reducing oxidative damage [25]. WD40 and cytochrome P450 are specifically related to ethylene and methyl jasmonate signalling [15, 16].

Furthermore, what is striking is that genes that encode WD40 and cytochrome P450 mirror their expressions depending on whether they are elicited by ethylene or methyl jasmonate signals [15, 16]. The synchronised behaviour of these genes based on their expressions seems to determine close co-ordination due to the elicitor. Our results with *G. imbricata* suggest that the expression of one gene can become activated and repressed without the assistance of another one, but expression is also linked to different signals related to both cystocarp disclosure and development. In *G. imbricata*, this means that WD40 gene expression responds to the ethylene signal when cystocarps are still non-visible, whilst this gene expression increases in the presence of the first cystocarps after methyl jasmonate treatment (disclosure period). Otherwise cytochrome P450 is expressed in the presence of the first cystocarps (developing cystocarps) when they are treated with ethylene. Conversely, after the methyl jasmonate elicitor, cytochrome P450 expression responds when cystocarps are still invisible. In both cases, as the cystocarps mature, expression holds over time without any significant changes between thalli with well-developed cystocarps and fully developed cystocarps [15, 16] (**Figure 3**).

In addition, the ascorbate peroxidase gene, which encodes a protein involved in the response to oxidative stress [6, 15], is also associated with the disclosure and development of cystocarps rather than with their maturity process [16].

Alternatively, polyamines, which are nonvolatile molecules but do have an important role in the process of maturing of the cystocarps, are synthesised through the candidate gene known as *ODC* [10, 14–16]. The synthesis of the polyamine precursor, putrescine, renders downstream spermidine and spermine due to the addition of one or two aminopropyl groups from decarboxylated SAM. Endogenous levels of these three polyamines—that is, putrescine, spermidine and spermine—are balanced by amine oxidase and polyamine oxidase, whilest H_2O_2 is released as a by-product of this reaction.

Monitoring amine oxidase gene, whose gene expression was seen to depend on the disclosure and development period of cystocarps, but also that once cystocarps have developed, reported that this gene expression would help to maintain polyamines levels (**Figure 3**) [16].

In short, our results confirm that genes encoding ROS proteins are related to physiological events. If we take the results as a whole, these behaviours of genes enable us to discern two action modes. Initially, WD40, cytochrome P450 and APX point to promoting the disclosure and development period of cystocarps, and they help to soften the thalli as up-expressions occurs. Meanwhile, amine oxidase expression shows a dual response. In other words, it helps cystocarp disclosure but it also balances ROS levels in order to fine-tune polyamine levels and prepare the thalli for the next time.

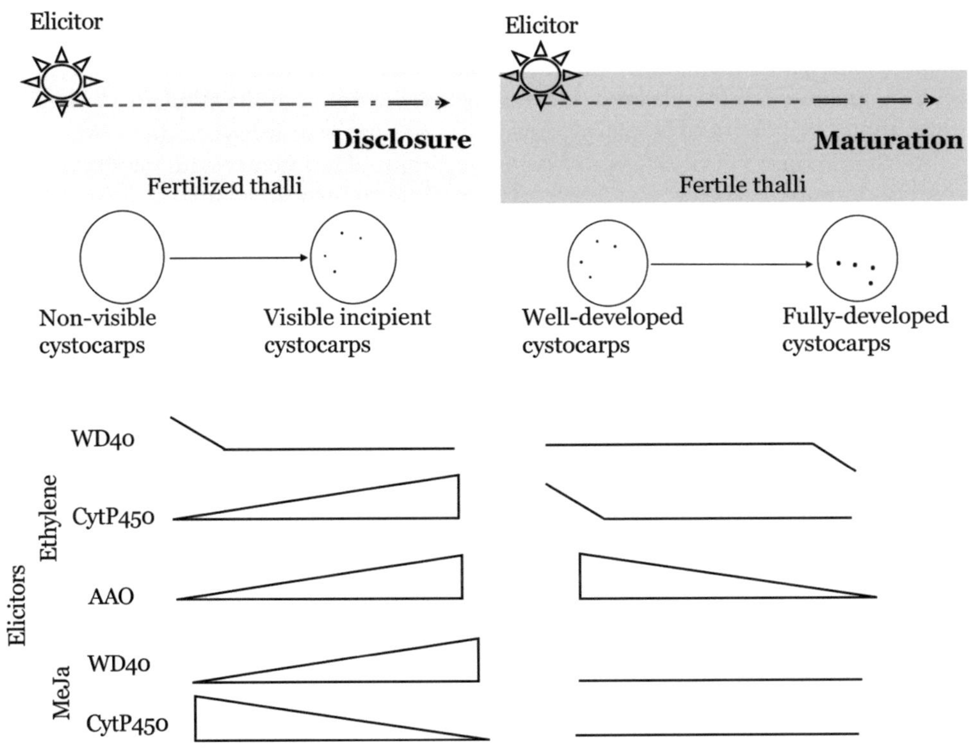

Figure 3.
Timeline of gene expression encoding stress proteins (WD40, cytochrome P450 and amine oxidase) for fertilised and fertile thalli of Grateloupia imbricata after ethylene and methyl jasmonate treatment. The sloping lines represent significant increase or decrease in gene expression with respect to absolute values (copies µl^{-1}). Horizontal line indicates no changes in gene expression relative to expression in control thalli. CytP450, cytochrome P450; AAO, amine oxidase; MeJa, methyl jasmonate.

4. Genes encoding proteins involved in biosynthesis pathways of growth volatile regulators

Despite the commercial importance of red seaweed, we still lack information on reproductive events if 'our' interest is to be able to control what happens over the course of the development and maturity processes of the reproductive structures and consequently manage to produce a large number of individuals.

Unlike the amount of information based mainly on next generation sequencing data, little progress has been made on the temporal control of genes, which affect growth and development. These aspects are of critical importance from the point of view of farming them. It is worth to highlight seaweeds that had received little attention worldwide to elucidate gene functions and to delve into the development and progress of functional genomic. Particularly in this section and as a practical goal, it is expected that molecular mechanisms related to volatile biosynthesis during carposporogenesis will provide tools for control and regulation of growth and developmental process in seaweeds. Thus, insight might allow to initiate a genetic programme for macroalgae which is economically valuable, increasing its viability and value.

The molecular nature of the signal(s) that control development and maturity of cystocarps is unknown, although efforts have been made in recent years to accurately describe the elicitation and disclosure periods of cystocarps in the red alga *G. imbricata*. One of the most striking features is that alterations in gene expressions even start prior to the presence of visible cystocarps, which seems to suggest that communication through signal pathways is essential for the disclosure of cystocarps.

Our research team has focused on gene screening related to proteins specifically involved in biosynthesis pathways of volatile growth regulators instead of profiles of up- and downregulated genes reported in massive sequencing. Although it is obvious that any attempt is appropriate given the lack of molecular information in red seaweeds, we ought to bear in mind the existence of environmental acclimation of algae and the tremendous changes in the levels of expression of a large number of genes during the disclosure, development and maturity of cystocarps. Incidentally, we have to remember that factors such as salinity and sporulation are connected, and our aim is to be able to discern precisely what is happening.

Hence in order to gain a better and more accurate insight into the control mechanisms underlying the reproduction of red seaweeds, the monitoring of specific genes, that in turn are also related to growth regulators and their biosynthesis, has been successful (**Figure 4**). In particular, gene-encoding enzymes needed for the synthesis of ethylene, such as SAM synthase (SAMS) and ACC synthase (1-aminocyclopropane-1-carboxylate synthase), genes that encode proteins of polyamine metabolism (spermidine synthase (Spd synthase); amine oxidase), genes encoding proteins of methyl jasmonate synthesis, such as jasmonic acid carboxyl methyltransferase (JMT) and putative methyltransferase (MT); and a gene that encodes a transcription factor involved in controlling responses to stress, growth and development (MYB, [26]), have been monitored. These gene expressions have provided valuable information and helped to shed light on the complex process of red seaweed reproduction. As for genes related to ethylene biosynthesis, these are directly involved in cystocarp development, that is, SAMS, Spd synthase and ACC synthase. Otherwise, all genes studied in relation to methyl jasmonate are indiscriminately induced in the absence of cystocarps (**Figure 5**).

In general, we can indicate that methyl jasmonate and ethylene signalling occurs either immediately after the elicitation period or during the disclosure period, respectively (**Figure 5**). The time course of different gene expressions indicates a temporal regulation of algal reproduction. As part of this temporal regulation, the

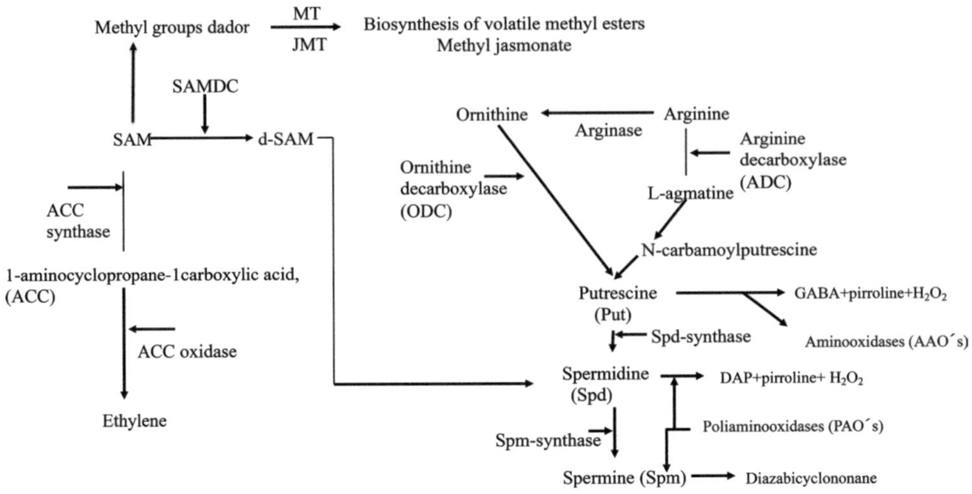

Figure 4.
Biosynthetic pathway for polyamines and connections with the pathways for the biosynthesis of ethylene and jasmonate. SAMS, S-adenosyl methionine synthase; d-SAM, decarboxylated SAM; MT, putative methyl transferase; JMT, jasmonic acid carboxyl methyl transferase.

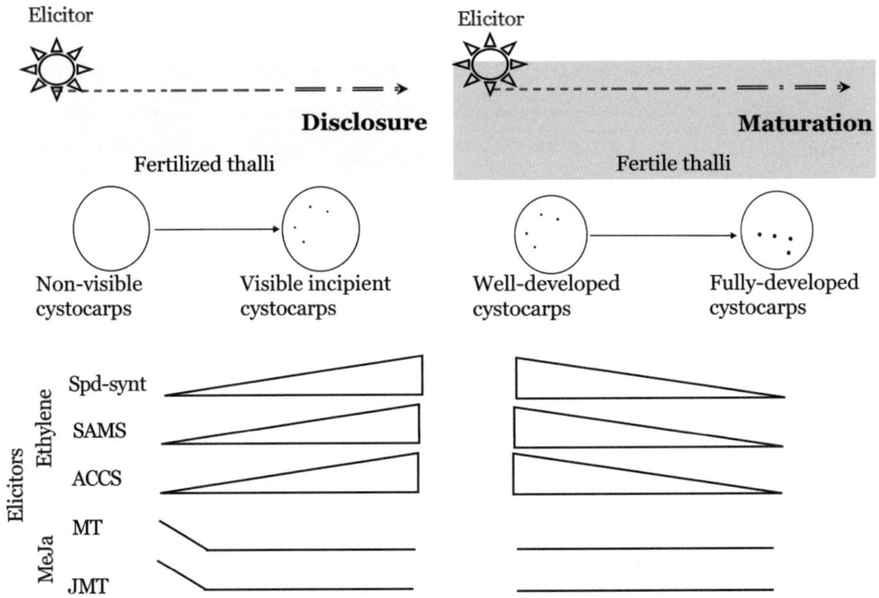

Figure 5.
Timeline of gene expression encoding biosynthesis proteins of ethylene (Spd synt, spermidine synthase; SAMS, S-adenosyl methionine synthase; ACCS, 1-aminocyclopropane-1-carboxylate synthase) and methyl jasmonate (JMT, jasmonic acid carboxyl methyl transferase; MT, methyl transferase) for fertilised and fertile thalli of Grateloupia imbricata after ethylene and methyl jasmonate treatment. The sloping lines represent significant increase or decrease in gene expression with respect to absolute values (copies μl^{-1}). Horizontal line indicates no changes in gene expression relative to expression in control thalli.

differential gene expressions represent the ability of seaweeds to sense ethylene and methyl jasmonate separately [15, 16].

Signal transduction—like the presence of cystocarps—brings up the question of whether the sensing of both volatiles could be elicited in accordance with the signal strength. The latter is within the bounds of possibility since (i) ethylene, which is the smallest volatile molecule, can easily cross through cell membranes and (ii) the hypothetical model of ethylene receptor for algae is a simpler structure than the one reported in higher plants. A priori, although both volatiles require membrane receptors, the fact is that the ethylene signal of the candidate gene elicited 12-fold the expression of methyl jasmonate despite the period where gene expression is reported [14]. This could be important for the fine regulation of disclosure and development of cystocarps.

To make this more difficult, something else caught our attention. We have also wondered whether signal strength can be interpreted as a differential response between elicitor signal and signal transduction. Signal transduction is assumed to be the responsibility of a complex and integrated molecular network. The network for one or another volatile could overlap in such a manner that this overlapping simplifies signal channelling. Contrary to what some may think, we do not rule out separate signalling networks. Nevertheless, there could also be a signal output modulation 'mechanism' that regulates the disclosure and development of cystocarps [16]. We are a long way from knowing what is happening—in other words, the differential perception of volatiles, the separate and overlapping signal pathways and signal strength. Nonetheless, we realise that gene knockout studies will be advantageous to confirm these issues. Although we have accomplished the primary goal of revealing the molecular mechanisms underlying red seaweed reproduction, further studies are required to identify and explore other factors involved in the regulation of gene expression.

5. Conclusions

This chapter has summarised our insight into the complexity of gene regulation during red seaweed reproduction. There are grounds to believe that temporal patterns of gene expression are orchestrated under the control of volatile growth regulators signalling during the disclosure and development of cystocarps. Progress is being made in understanding how thalli transduce these volatile signals.

Acknowledgements

This work has been financed with funding from the Universidad de Las Palmas de Gran Canaria. Funding from Consejería de Economía, Industria, Comercio y Conocimiento del Gobierno de Canarias to PGJ is acknowledged (PROID2017010043 ACIISI; CEI2018-20 ULPGC).

Conflict of interest

The authors declare that there is no conflict of interest.

Author details

Pilar Garcia-Jimenez* and Rafael R. Robaina
Department of Biology, Faculty of Marine Sciences, Universidad de Las Palmas de Gran Canaria, Las Palmas de Gran Canaria, Canary Islands, Spain

*Address all correspondence to: pilar.garcia@ulpgc.es

References

[1] Kloareg B, Quatrano RS. Structure of the cell wall of marine algae and ecophysiological functions of the matrix polysaccharides. Oceanography and Marine Biology: An Annual Review. 1998;**26**:259-315

[2] Schopfer P. Hydroxyl radical-induced cell-wall loosening in vitro and in vivo: Implications for the control of elongation growth. The Plant Journal. 2001;**28**:679-688

[3] Fry SC. Oxidative scission of plant cell wall polysaccharides by ascorbate-induced hydroxyl radicals. The Biochemical Journal. 1998;**332**:507-515

[4] Guzman-Uriostegui A, Garcia-Jimenez P, Marian F, Robaina RR. Polyamines influence maturation in reproductive structures of *Gracilaria cornea* (Gracilariales, Rhodophyta). Journal of Phycology. 2002;**38**:1169-1175

[5] Garcia-Jimenez P, Brito-Romano O, Robaina RR. Occurrence of jasmonates during cystocarp development in the red alga *Grateloupia imbricata*. Journal of Phycology. 2016;**52**:1085-1093. DOI: 10.1111/jpy.12467

[6] Garcia-Jimenez P, Robaina RR. Volatiles in the aquatic marine ecosystem: Ethylene and related plant hormones and sporulation in red seaweeds. In: Kumar M, Ralph P, editors. Systems Biology of Marine Ecosystems. 1st ed. Gewerbestrasse, Switzerland: Springer International Publishing; 2017. pp. 99-116. DOI: 10.1007/978-3-319-62094-7_5

[7] Kitade Y, Asamizu E, Fukuda S, Nakajima M, Ootsuka S, Endo H, et al. Identification of genes preferentially expressed during asexual sporulation in *Porphyra yezoensis* gametophytes (Bangiales, Rhodophyta). Journal of Phycology. 2008;**44**(1):113-123

[8] Shen S, Zhang G, Li Y, Wang L, Xu P, Yi L. Comparison of RNA expression profiles on generations of *Porphyra yezoensis* (Rhodophyta), based on suppression subtractive hybridization (SSH). BMC Research Notes. 2011;**4**:428

[9] Uji T, Monma R, Mizuta H, Saga N. Molecular characterization and expression analysis of two Na^+/H^+ antiporter genes in the marine red alga *Porphyra yezoensis*. Molecular Biology Reports. 2012;**39**(8):7973-7980

[10] Garcia-Jimenez P, Rodrigo M, Robaina RR. Influence of plant growth regulators, polyamines and glycerol interaction on growth and morphogenesis of carposporelings of Grateloupia cultured in vitro. Journal of Applied Phycology. 1998;**10**:95-100

[11] Sacramento AT, Garcia-Jimenez P, Alcazar R, Tiburcio AF, Robaina RR. Influence of polyamines on the sporulation of Grateloupia (Halymeniaceae, Rhodophyta). Journal of Phycology. 2004;**40**:887-894

[12] Sacramento AT, Garcia-Jimenez P, Robaina RR. The polyamine spermine induces cystocarp development in the seaweed Grateloupia (Rhodophyta). Plant Growth Regulation. 2007;**53**:147-154

[13] Garcia-Jimenez P, García-Maroto F, Garrido-Cárdenas JA, Ferrándiz C, Robaina RR. Differential expression of the ornithine decarboxylase gene during carposporogenesis in the thallus of the red seaweed *Grateloupia imbricata* (Halymeniaceae). Journal of Plant Physiology. 2009;**166**:1745-1754

[14] Montero-Fernandez M, Robaina RR, Garcia-Jimenez P. In silico characterization of DNA motifs associated with the differential expression of the ornithine decarboxylase gene during in vitro

cystocarp development in the red seaweed *Grateloupia imbricata*. Journal of Plant Physiology. 2016;**195**:31-38. DOI: 10.1016/j.jplph.2016.03.004

[15] Garcia-Jimenez P, Montero-Fernandez M, Robaina RR. Molecular mechanisms underlying *Grateloupia imbricata* (Rhodophyta) carposporogenesis induced by methyl jasmonate. Journal of Phycology. 2017;**53**:1340-1344. DOI: 10.1111/jpy.12594

[16] Garcia-Jimenez P, Montero-Fernandez M, Robaina RR. Analysis of ethylene-induced gene regulation during carposporogenesis in the red seaweed *Grateloupia imbricata* (Rhodophyta). Journal of Phycology. 2018;**54**:681-689. DOI: 10.1111/jpy.12762

[17] Tripathy BC, Oelmüller R. Reactive oxygen species generation and signaling in plants. Plant Signaling and Behavior. 2012;**7**(12):1621-1633

[18] Mittler R. ROS are good. Trends in Plant Science. 2017;**22**(1):11-19

[19] Wuosmaa AM, Hager LP. Methyl chloride transferase: A carbocation route for biosynthesis of halometabolites. Science. 1990;**129**:160-162

[20] Roje S. S-Adenosyl-L-methionine: Beyond the universal methyl group donor. Phytochemistry. 2006;**67**:1686-1698

[21] Garcia Jimenez P, Robaina RR. Effects of ethylene on tetrasporogenesis in *Pterocladiella capillacea* (Rhodophyta). Journal of Phycology. 2012;**48**:710-715. DOI: 10.1111/j.1529-8817.2012.01156.x

[22] Miller G, Suzuki N, Ciftci-Yilmaz S, Mittler R. Reactive oxygen species homeostasis and signaling during drought and salinity stresses. Plant, Cell and Environment. 2010;**33**:453-467

[23] Weinberger F, Lion U, Delage L, Kloareg B, Potin P, Beltran J, et al. Up-regulation of lipoxygenase, phospholipase, and oxylipin-production in the induced chemical defense of the red alga *Gracilaria chilensis* against epiphytes. Journal of Chemical Ecology. 2011;**33**(7):677-686

[24] Garcia-Jimenez P, Brito-Romano O, Robaina RR. Production of volatiles by the red seaweed *Gelidium arbuscula* (Rhodophyta): Emission of ethylene and dimethyl sulfide. Journal of Phycology. 2013;**49**:661-669. DOI: 10.1111/jpy.12083

[25] Xu C, Min J. Structure and function of WD 40 domain proteins. Protein and Cell. 2011;**2**:202-214

[26] Ambawat S, Sharma P, Yadav NR, Yadav RC. MYB transcription factor genes as regulators for plant responses: An overview. Physiology and Molecular Biology of Plants. 2013;**19**:307-321